SPUR PUBLICATIONS

POULTRY FANCIERS' LIBRARY

General Editors:

DR. J. BATTY   MRS. M. BATTY

BANTAMS AND MINIATURE FOWL

OTHER BOOKS AVAILABLE

Understanding Old English Game—Large and Bantams
Dr. J. Batty

Royal Pastime of Cockfighting
Robert Howlett
(Reprint of 1709 classic)

Poultry Culture for Profit
Rev. T. W. Sturges

Managing Poultry for Exhibition
H. Easom Smith

Frontispiece: BANTAMS: MALE AND FEMALE

(From an oil painting by Gwenllian Woods)

Top

Ancona Andalusian Araucanas (Lavender)

Bottom

Australorp Barbu d'Anvers (Quail) Barnevelder (Double Laced)

Bantams
and Miniature Fowl

W. H. SILK

THIRD EDITION

Revised by Dr. J. Batty
assisted by other Poultry Fanciers

PUBLISHED BY THE SPUR PUBLICATIONS COMPANY
Hill Brow, LISS, Hampshire, GU33 7PU

ISBN 0 904558 14 2

SECOND EDITION 1974
THIRD EDITION JANUARY 1976
REPRINTED DECEMBER 1976

Printed in Great Britain by
REDWOOD BURN LIMITED
Trowbridge & Esher

CONTENTS

ILLUSTRATIONS

EDITORS' FOREWORD

IN REVISING a book written by an acknowledged authority it is desirable to keep the best features, amending only when necessary. This in fact is what has been done. However, even this approach has necessitated considerable work. Inevitably, after 20 years many changes have taken place in the character of the birds. With some breeds there has been considerable improvement in type, colour and other features.

Thanks are offered to the fanciers who provided photographs or blocks for reproduction of their prize winning birds. In addition, due acknowledgement is given to the invaluable assistance given by a number of these fanciers in providing notes and hints on the various breeds they keep. Nobody can know all the finer points of every breed and for this reason expert advice was much appreciated. A List of Acknowledgements is given on the next page. Reference was made to *British Poultry Standards* and various Poultry Club Year Books as well as the *Fur and Feather*.

We were unable to include as many photographs as we would have liked. Inevitably some had to be excluded, including our own birds.

J. Batty

M. Batty

ACKNOWLEDGEMENTS

Grateful thanks are offered to Fanciers who have given advice and supplied photographs. The principal of these are shown below.

Miss Caroline Batty.
G. Edwards.
J. C. Duckett.
A. R. Kerr.
M. L. Goodfellow.
Alex King.
E. C. Ellis.
Miss Veronica Mayhew (C. and V. Mayhew).
Mr. and Mrs. W. B. Johnson.
W. J. Kemp.
A. Stafford.
J. D. Kay.
C. A. Parker.
J. L. Milner.
T. F. Bower.
W. G. Groucott.
Mrs. W. Roxburgh.
A. Howard.
Ken Binns.
Raymond Shaw.
Messrs. Jas. R. and M. Smith.
P. Parris.
Alan Maskrey.
Will Burdett.
J. Shortland.
A. Anderson.
J. Brannan.
J. Hey.
R. Nottrodt.
W. Langton.
Robin McEwan.
Norman Parkinson.
D. W. Ledward
For details of Houses and Runs:
    Park Lines and Co.
    Harry Hebditch Limited.

CHAPTER 1

MAKING A START

DEFINING A BANTAM

A BANTAM is a diminutive fowl which is:

1. naturally small and has no large counterpart; e.g. Belgian Bearded, Booted, Sebright, Japanese, Nankin, Pekin and Rosecomb; or
2. is man-made being bred down from the large breed.

Most of the "created" bantams have been brought down to size by crossing the large breed with some other bantam. The Sebright has no large equivalent being developed as a distinct breed by the late Sir John Sebright. Game bantams have been known for generations and according to one of the earliest books should be regarded "as occupying a distinct place in the family" of bantams*.

In terms of size bantams are usually reckoned to be approximately a quarter of the size of the equivalent large breed.

ADVANTAGES OF BANTAMS

In reasonably good circumstances breeding and exhibiting are economical and profitable, providing many hours of pleasure at a cost much less than keeping large fowl. Expenditure on food is indeed surprisingly little. The number of birds that may be kept mainly on scraps from a normal household, augmented by a very little meal (and grain when available) astonishes most people. Scraps and offal from a household of five will keep a dozen bantams healthy, and keep them laying.

Space necessary for breeding is very small. Even a backyard sometimes proves sufficient when a garden is not available; but ample light and air are essential.

*Wingfield W. and Johnson C. W., *The Poultry Book,* Wm. S. Orr, London 1853, p. 195.

Exhibiting is cheaper. Travelling hampers, training pens and appliances are all much less costly than with large poultry, while bantam breeders definitely don't have to bother about high freights—they can send four exhibits to show at the cost of one large bird. The housewife seldom refuses to take charge of such charming pets—a great advantage in short winter months, when menfolk are away all the daylight hours.

Finally, if the right breed be selected and sound methods used, eggs laid are out of all proportion to body size, both in weight and numbers. Many breeds will provide the household with $1\frac{1}{2}$ oz. eggs and plenty of them. Higher weights should not be expected, though they are by no means rare.

Different breeds sometimes require different housing. The house that suits hardy breeds like Old English Game or Sussex will not suit docile, comparatively inactive Pekins, Japanese or Polands. When you have read the chapters dealing with breeds you will know why.

SELECTING THE APPROPRIATE BREED

Before preparing your accommodation, therefore, visit shows and examine closely the various breeds. Don't despise exhibitions, even though you want your birds mainly as layers; remember shows are the shop-windows from which you, and others like you, make your choice.

So choose your breed first; and get a bantam breeder to help you. If you intend to exhibit, don't select, to commence with, a variety that is difficult and highly-developed in show points. Select a simpler variety, and when experience is gained make a change to the more purely fancy breeds. Any good fancier will help.

Such breeds as Wyandottes, Sussex and Rhode Island Reds are much less difficult than delicately-laced Sebrights or highly-developed Rosecombs. This does not mean that any variety can be produced to perfection with ease; but some are more suitable for beginners.

After deciding your breed, get sound advice and prepare your housing. If ordinary stock designs are suitable for the breed selected, don't bother to make houses—buy them. In normal times they can be bought too cheaply to make home-construction worth while. Read the later chapter on housing, then decide this point; but if your choice lies in purely fancy breeds like feather-legged Cochins or Pekins, crested Polands or low-carriaged Japanese, you should consider making your own. A little thought in advance will save later disappointment and work.

When preparing your lay-out, spare a thought for your neighbours and for local amenities. Bantam runs can be attractive so that they blend with the surrounding garden.

2

EXHIBITION OR UTILITY?

Now comes a very important point. If you only want pets, and are not particular about exhibition character, you needn't study this paragraph further. Just get a knowledgeable friend to buy you some sound healthy stock and rest content—remembering only that you might just as well have pure-bred stock. Your surplus young birds will then command ready sales at remunerative prices. It costs no more to keep good birds than bad. But if you want the best, or desire to exhibit, take some trouble to get a good start.

You have two alternatives—you can buy breeding stock or hatching eggs; and of these the former is much the better. Make up your mind what you can pay, and get a good *breeder* (not a dealer) to provide properly-mated birds to that value. For his reputation's sake any good fancier will supply good stock; but don't expect a high class pen for next to nothing. A man who has spent years of his life at the job can't be expected to sell his best, or give you the benefit of his work and knowledge, for a mere song.

BUYING STOCK BIRDS

Good stock birds are usually sold in trios, and will probably cost an average of £4-£5 each. Birds at this price would not necessarily be top show birds; but you don't want winners for breeding. Contrary to general belief, high quality breeders are of more value than show birds, and for the best high prices are often paid. Remember in the long run it pays to acquire the best stock—even if it costs more!

As show birds are not usually best for breeding, don't buy winners for stock purposes. The sound fancier values the birds that produce his winners much more than the winners themselves. That is why you should not expect your initial breeding stock to be bought cheaply.

BUYING EGGS

If you buy eggs for sitting you may possibly make a less expensive start, but don't stint the price. A good breeder just won't sell eggs cheaply—why should he? They are valuable, and may produce chicks that will defeat him at shows. Usually also he agrees to replace infertile eggs. (This does not mean all eggs that fail to hatch—it means eggs not containing the germ from which a chick may grow.)

Egg boxes, carriage and replacements will also add to the cost, so it is useless to expect the best eggs for nothing. Normal price is now around £3 per dozen; but at certain seasons the beginner may buy at reduced prices.

3

If you buy breeding stock, get them in the Autumn. They will cost you less than if (like most beginners) you wait till spring, when breeders have carried their birds through the winter months and perhaps severely culled them. If, however, you buy hatching eggs, get them in the Spring. Bantams are most successfully hatched from April to the end of May—rather later than large breeds.

FUTURE BREEDING

The golden rule in breeding is to hatch as many chicks as possible from your *best* birds—which will usually be few; and having reared them, select your females very rigidly for future breeding. You will then, if you are wise, mate a cockerel to hens the next year, possibly sending to the original source for a male of the same strain but not closely related. However, it may be that amongst your own chicks are two or three obviously sturdy suitable cockerels, so you must be guided by events.

Fairly close in-breeding (line breeding) is almost invariably practised in bantams, particularly in highly-developed show stock; so you may perhaps wait a year or so before purchasing another male.

Either alternative, carried out sensibly and carefully, will set you on the road to success; and from then on your achievements will only be limited by your knowledge and your flair for the game.

ESSENTIAL THREE "Cs"

If birds are to be of maximum usefulness good, fit stock is essential. Fanciers should look for:
1. Condition—overall fitness and sound plumage.
2. Colour—complying with standard (critical in some breeds).
3. Conforming to type.

Remember it costs no more to keep top-class birds than it does poor specimens.

CHAPTER 2

CHOOSING YOUR BREED

SELECT BIRDS YOU FANCY

ALMOST INVARIABLY the first thing beginners ask is advice on what breeds to keep. To them all there can be only one reply—visit shows and make your own choice. You will also be well advised, when inspecting the various exhibits, to ask well-known judges, breeders or showmen the advantages and difficulties of any variety that pleases your eye.

There is almost unlimited choice—and great charm in them all. The most exciting, of course, are the difficult ornamental breeds, which you would do well to avoid until you have gained experience. Nevertheless, the beginner is likely eventually to do best with the variety he most admires, so long as he sticks to it and is not easily disheartened by early set-backs.

All the more fancy breeds, however, are not difficult. Some of the old breeds, possibly crested and feather-legged, such as Polish, Pekins and Belgians, breed very true if you start with good stock; but don't choose varieties that require very heavy show preparation. Above all, don't follow the usual beginner's practice of keeping too many kinds.

One breed will do to start with. Don't add a second for a year or so, when you will have gained experience with the first. This advice will almost certainly fall on deaf ears, but it is the best in the world.

Your choice of breeds must depend entirely upon your own personal outlook. If you are chiefly interested in showing you will select from genuine fancy breeds, which are mostly very small, highly inbred, and of ancient lineage. If your main desire is for a useful hobby that will provide the household with eggs you will choose a more modern variety, larger, more vigorous and bred for production; and if you merely want pets to run about the lawn your choice is almost unlimited, from that dashing cavalier the Old English Game Cock through all types of soft-feathered, motherly but tiny breeds, all with their own claims to charm.

5

HARD AND SOFT FEATHER

Under Poultry Club rules bantams are normally divided into two main groups which explain themselves—hard feather and soft feather breeds. Sometimes this is varied a little by splitting them into Game and Variety bantams. The two groupings are not quite alike. All game birds are hard-feathered, but not all hard-feathered birds are classed as game.

This gives rise to a few peculiarities. Indian Game, Jubilee Game, Sumatra Game, Aseels, and Malays may be shown in the hard feather section, but must not be shown in game classes in spite of their names. Where no classification is reserved specifically for "hard feather" (as distinct from "Game") they must be shown in the Variety bantam section, but are never eligible to compete in a "soft feather" group or class.

BANTAMS AS LAYERS

If you want eggs before charm, try Sussex, Rhode Island Reds and Anconas. If you want charm with perhaps as many eggs, you might keep Minorcas, Leghorns, and Buff Rocks.

You can't expect eggs throughout the year though, unless you accept a body-size too large for exhibition; and it is useless expecting to buy flocks of laying bantams. You must buy breeding stock and produce your own. If you are satisfied however with oversized birds, and if you expect lots of eggs $1\frac{3}{4}$ oz. in weight, you shouldn't bother with bantams—you should keep large breeds.

The larger and more vigorous breeds will pay their way, but they won't lay 200 eggs a year, they won't lay throughout severe winter weather, and they should not be expected to produce eggs heavier than a maximum of $1\frac{1}{2}$ oz. each. To pretend the contrary is false and misleading.

Good bantams should weigh not more than one-fourth the weight of their comparable large breeds. Why then expect them to lay eggs almost equal in size as well as in numbers?

UTILITY BREEDS

In addition to "fancy" (or ornamental) breeds and sturdy layers, there is a large group, of all-round excellence—birds that possess exhibition charm combined with ability to produce eggs in plenty. Pre-eminent amongst these is probably the Wyandotte, which exists in multitudes of colour-varieties, some of which are almost unbeatable layers—mainly of eggs weighing about $1\frac{1}{4}$ oz.

Barred Plymouth Rocks are also excellent. Hamburghs will lay plenty of large eggs, and Orpingtons (with Australorps, their alleged close relations) should not be overlooked.

SMALL BANTAMS AS LAYERS

From the smallest breeds like Sebrights and Rosecombs, which are standardised at weights very little more than one pound, you mustn't expect eggs of much more than an ounce each—though one of our most highly-developed bantams, the Modern Game, lays eggs so big in proportion to its size, that it can't be expected to lay them without risk.

HOW THEY WERE CREATED

Few of our bantam breeds have any appreciable proportion of large breed blood in them—some in fact have none. That is why breeds like Welsummers and Barnevelders lost their deep brown eggs for which the large varieties are noted.

A complete answer to those who claim that their bantams have been bred down pure from large breeds (without the aid of a bantam cross) is found in the last sentence of the previous paragraph. Originally no single breed of bantam carried the deep brown eggs factor because it was lost in process of bantamising, purely because the generic egg-colour in all bantams is white or pale cream.†

Bantamising a big breed is not difficult; but it just can't be done under a lifetime unless you introduce bantam blood. Those who like to improve type in bantam breeds often use a large breed outcross for that purpose. It cannot be denied that type in bantamised editions of large breeds is often so poor as to make such a cross advisable.

Bantams do not have strong commercial possibilities. Even in the larger and sturdier varieties the factors that contribute to commercial success (sex-linkage, auto-sexing and sexing by physical examination) are practically unknown; while in genuine bantam breeds hatchability, rearability and productivity are markedly inferior to large breeds. On a utility farm two large breed chicks or more could easily be reared for every bantam chick possible to produce.

Bantams are, therefore, birds for the fancier; the enthusiast who breeds out of interest and who takes a delight in effecting improvements. All have charm; some are good layers, whereas others produce only a few eggs each year. Many are capable of laying much better than is commonly imagined.

---

† A few breeders now appear to be getting brown eggs, but this does not invalidate the argument.

CHAPTER 3

HARD FEATHER BREEDS

As PREVIOUSLY NOTED, these are in two groups, one comprising Modern and Old English Game (the only breeds eligible for showing as game), the other including Aseels, Malays, Sumatra Game, Indian Game and Jubilee Indian Game. The latter group must not be shown in game classes, but are eligible for classes specified "hard feather", and therefore must not be shown in soft feather classes.

Where no special classes are provided and where no breed classification is given, this group can only be shown in classes reserved for A.O.V. (Any Other Variety).

Jubilee Indian Game may not be shown in classes reserved for Indian Game but a combined class may be possible.

Because our two genuine game breeds (Modern and Old English Game) comprise by far the largest section of the hard-feathered varieties they must come first in this chapter.

MODERN GAME

Modern Game are entirely a fancier's breed, developed in a surprisingly short space of time from birds of normal build into sleek long-limbed birds which nevertheless have a strangely fascinating charm. Many problems in colour-breeding are found in both Game breeds; and therefore this chapter, dealing primarily with Moderns, is largely just as applicable to Old English.

THE STANDARD

In Moderns **shape** and **style** and **colour** are of vital importance. The male should be tall, reachy and graceful, with prominent square shoulders. A rounded or roached back is a very serious fault. A short flat back tapering towards the stern is essential.

Shanks must be fine, round and very long, and thighs should be long and muscular, set well apart. Flat shins are a severe exhibition fault and are strongly hereditary. The hind toe should be as nearly as possible direct in line with the middle toe. Avoid

Pair of
Black Reds

Pair of
Birchen

Pair of
Brown Reds

Pair of
Golden Duckwing

Figure 1. Modern Game in Various Colours

duck feet and back toes that do not reach the ground—though in very stylish birds the back toe is often carried too high. This is a fault, but not as serious as carrying it sideways or close to the foot, which is severely penalised and is inclined to be hereditary.

Wings are required short and curved, just long enough to meet at stern, fitting close to the body and carried well up, but avoiding goose wings—which means carrying them partly across the back.

There should be a long, snaky head, with eyes large and prominent—not at all like the fighting head of an Old English Game cock. Eyes in Brown-Reds and Birchens should be as dark as possible—black if you can get it. In other colours they should be bright red. The neck should be fine and long, with close-fitting hackles.

Tail and tail carriage are particularly important. Main tail feathers should be narrow, fine, and close-fitting. Male sickles should reach only a couple of inches or so beyond the tail proper, and are slightly curved and close together, not forked. Tails should be carried only slightly above body line.

In body-build, avoid slab sides due to lack of curve in wings, and don't breed from turkey-breasted birds or your stock will acquire ugly, prominent keels.

We are dealing with highly-bred stock, so size is very important. Most old-time fanciers aimed at the smallest specimens for show. These were seldom good stock birds, and indeed females often died when passing their first eggs. Nowadays we are less obsessed with smallness, and allow a certain latitude above standard weights, which are: Cock 20–22 oz., Hen 16–18 oz.

In considering colours, attention should be paid to body, legs and eye colours: they should be in accordance with the standard as well as "matching".

Novices often fail through buying show birds and breeding from them, producing weeds that are difficult to rear and that lack game character. Don't use females of less than 16 oz. for breeding; and 18 oz. to 20 oz. will be more satisfactory for males. Higher weights will usually produce better stock, and the worst chicks to rear are bred from birds that regularly go the rounds of shows.

COLOUR REQUIREMENTS

Black-Reds.—Males: Face, lobes, head, comb and wattles (what remains of them after dubbing) must be bright red, with light orange neck hackle shading to yellow and free from striping. Saddle hackle should match the neck; back and wing-bow should be rich crimson.

10

Wing-ends or bays must be light chestnut, wing-bars glossy steel-blue, wing-butts black. Breast and thighs black, free from lacing, ticking or other markings; but adult cocks, as they age, will seldom be perfectly clear-breasted.

Sickles and tail should be black throughout, without red shafts or fringing (found in pullet-breeders and a fault in showing). Eyes should be red and beak dark green; legs should be willow.

Females should be as described for males in head points. It is, however, sometimes difficult entirely to avoid white in lobes. Doubtless this fault would be more easily eradicated if showmen refrained from using scissors; and while dubbing is permitted in males this is difficult to penalise. Sometimes the sand-paper on a matchbox serves the same purpose and needs less skill; but both methods should be penalised.

Neck hackle should be pale gold with pale lemon edging round narrow black central striping; throat a pale salmon, breast a richer salmon shading to a pale tint at thighs and under-parts.

Body and wings are brownish drab, soft and even, and very finely pencilled with black; the desired colour being a light partridge-brown with a slight golden tinge, free from rust or ruddiness. Coarse pencilling is objectionable in the show pen, though sometimes useful in cockerel-breeding.

Flight feathers are best unpencilled, and tail should be black, except coverts and top feathers, which match body colour.

BREEDING BLACK-REDS

Use only healthy stock and breed only from your best birds. It is better to use two pens, one mated to produce good cockerels and one to produce show pullets.

For the cockerel-breeding pen, use the brighest-coloured, tallest cockerel, with prominent shoulders and short back, mated to say three reachy close-feathered hens, very pale in hackle colour and crown. Don't use hens with dark caps and deep-coloured hackles, which should be as pale as possible right to the head. Body-colour and wings should incline to foxiness or reddish tinge, with rich gold fringes round each feather. Select females short in back, compact in body, prominent in shoulders and short in tail with plenty of reach and style.

Wheaten blood was once used regularly to produce bright top colour, but not often nowadays. Too few breeders can spare time and accommodation for the necessary extra breeding operations, because progeny must be clearly marked.

In pullet-breeding, use a male known to be from a pullet-breeding strain. He will be darkish in colour, with brickish top-colour, as even as possible in shade from cap and hackle to tail. Wing and shoulders should be black and free from lacing. Adult cocks will probably be laced on breast and weak in wing-bays. Sickle shafts will be red instead of black.

11

Mate with him about three of the very best hens you have—sound in colour, free from coarse pencilling and clear of rust or foxiness. You can't produce exhibition females from hens of faulty colour.

It goes without saying that your females must all be short-backed, of good shape, reach and style; but they need not be the smallest birds.

Where the novice has not space or accommodation for two pens, he must make the best of breeding from one; so he should either content himself with breeding for good specimens of one sex only, or adopt a compromise by trying one pen to produce a moderate proportion of good birds in both sexes. He will produce a number of wasters, but he must put up with that.

The head of the pen should then be a cock of pullet-breeding character but as light in top-colour as possible—the paler the better. Mate to him, say, two cockerel-breeding females and two of standard exhibition colour. There is then a sporting chance of reasonably good success.

There is no way of avoiding in-breeding in Moderns—or indeed in any high-class stock; so the second season it is wise to pick out the best of your pullets and mate them back to their sire, following similar methods with your best cockerels and the parent hens. Unless you do this you cannot build up a recognisable strain of your own.

Piles.—*Males:* The most handsome of Moderns are undoubtedly Pile cockerels. They are in the nature of a semi-albino edition of Black-Reds. The cockerel should be like the exhibition Black-Red, except that he must be sound clear white where the Black-Red is black—namely, breast, wing-butts, wing bars, thighs and tail; but legs and beak *must be* rich orange yellow. Willow legs are simply not considered. Head and face should be as already described, with eye as ruby red as possible.

A perfectly white breast is important, but for show it does not greatly matter whether colour is dark and even throughout in wings, back and hackles, or whether top colour is bright with rich orange hackles. Judges differ, but agree that wing-ends must be dark.

Females should have rich salmon breasts shading off to thighs and underparts. Body, tail and wings should be pale creamy-white, as clear of foxiness or red tinge as possible. Rich dark breasts and clear wings are difficult to obtain. You will breed twenty with rose wings or dark body colour for every one clear on breast and back. Foxy feathers are often removed from wings, but this should be discouraged.

Exhibition Pile cockerels are sometimes bred from Black-Reds, so when buying stock find out how your birds are bred.

For cockerel-breeding use a cock free from lacing or other colour on breast, perfectly sound in white and with sound wing-ends. Without good bays he is useless for producing good cockerels. Mate him to three or four tall pullets, deep in breast

Figure 2.   Miss Caroline Batty's Birchen Modern Game Female.
Winner of the Bilkey Cup

colour, reachy and stylish, short in back and with prominent shoulders, rosy on wing but not creamy on wing-ends. This pen will breed good show-type cockerels, but pullets will be warm in colour—not good for show but invaluable as breeders when mated back to their sire.

For pullet-breeding, use a male with top colour one uniform shade of darkish-brick, clear white in wing bars and shoulders, and coloured on breast if possible. Females should have good salmon breasts (not necessarily dark) and be free from cream or rosiness and clear in body colour. From this pen cockerels will be too pale in colour for show, but if good in bays and white ground will be useful for breeding.

13

Where space is too limited for double-mating, use the best-coloured show-type male you have, and mate him to, say, two females clear on wing and with deep salmon breasts, and two that are rose-winged. From such a pen you will get cockerels that are good all round, and pullets of good show quality but not clear on wing.

When breeding Piles pure for a number of years, top colour in cockerels is lost. Black-Red colour is then introduced, using a cockerel rich in top colour and saddle (the richer the better) sound black in wing bars, with black breast free from lacing or ticking. You won't get Piles with sound white breasts otherwise; and wing-ends must be dark and sound to tips.

If you mate this male to two or three pale pullets (pile-bred white or lemon-piles) you will breed pile cockerels of grand colour. Lemon-pile pullets are clear on wing and almost white on breast, and these are often shown, though non-standard. This method breeds yellow-legged males, but most of the females will be willow-legged and useless for show, though they might be used for breeding to a sound Pile-bred pullet-breeding cock.

Similarly, when Pile pullets come pale in breast mate them to a yellow-legged Pile-bred Black-Red cockerel, choosing one with even light-brick coloured top and wing bars and butts free from lacing. Lacing on breast, of course, is not harmful.

Brown-Reds.—*Males* have faces of very dark mulberry (gipsy faced)—the darker the better. If it approaches black it is ideal. Eyes must be really dark. If they look black so much the better. Beak, legs and feet are black, hackles pale lemon, saddle and back rich lemon, wing bars and tail black.

Breast must be black, with fine, delicate pale lemon lacing round each feather, lacing to start at throat and extend down the breast to a point just below shoulders. Lacing used to be required to run well down the thighs, but this is too coarse and heavy for modern taste.

Females are similar in eye, face and legs. Neck hackle is pale lemon from crown of head downwards, with narrow black striping. Dark caps must be avoided. Breast should be black, each feather from throat downwards laced with pale lemon; lacing uniform, fine and delicate, and not carried down much below shoulders. Body and wings are glossy black, free from shaftiness and lacing; and backs and shoulders should be clear of laced feathers. All these points are perhaps not easy in one bird, but they can be obtained. To be successful cockerels should have beautiful lemon top colour and females pale lemon hackles.

To produce show cockerels mate your best-coloured male, with plenty of pale lemon top-colour, darkest of eyes, and clear breast-lacing to females tight in feather, pale in neck hackle and with pale lemon caps. This is important. If they fail in possessing laced backs, don't worry—you'll get more top-colour on the cockerels they breed.

14

Pullets from this pen won't be any good for show, but if some of them are laced on backs, wings, and saddle keep them for cock-breeding, and next season mate them to their sire. You will soon learn how to use these double-mating and in-breeding methods if you are really interested.

For breeding show pullets a cockerel of darker shade, approximating to deep orange top colour instead of lemon, is used. To him mate females clear on back but well-laced on breast. These females should, in fact, be of exhibition standard, with bright pale hackles. Cockerels bred from this pen would be poor in colour but useful for pullet-breeding. It takes several seasons to build up your own recognisable strain.

If your Brown-Reds regularly come too dark in lemon colouring, you must introduce Birchen blood (see later description). Take care, though, that in doing so you don't lose gipsy faces and dark eyes. Any Birchen cock used needs good silvery-white top-colour and well-laced breast. Mated to Brown-Red females (which need not be the palest in hackle-colour) these should produce cockerels of really bright lemon colouring.

If you mate one of these the next season with the best pullets, and another with the parent hens, you will improve lemon colouring and retain good face and eye colour.

Birchens: These resemble Brown-Reds in all respects except that where Brown-Reds are lemon-coloured, Birchens are silvery-white—the purer and whiter the better consistent with retaining dark faces and eyes.

Figure 3. Birchen Modern Game Male (Miss Caroline Batty)

15

Birchen pullets have a tendency to be dark-capped. The breast should be delicately and distinctly laced from throat down to the level of shoulder-points. Too often they have a mere thumb-print of lacing or, conversely, excessive lacing.

Eye colour is not often as dark as in Brown-Reds, but red eyes should be penalised and should never be tolerated in the breeding pen. They should be mated for breeding as described for Brown-Reds; or they can be produced by mating a Silver Duckwing cockerel to well-laced Brown-Red females.

It is not possible fully to describe breeding methods for all colours in every breed. Beginners should experiment on the lines indicated and study results and methods for themselves.

Duckwings are of two main colours:
1. Golden.
2. Silver.

GOLDEN—*Males:* Neck and saddle hackle are silvery-white, as free from black striping as possible. Breast must be sound black, and wing butts, wing bars and thighs are also black. Tails are as described for Black-Reds. Saddles and wing-bows are rich yellow or orange, shading to silvery-white in saddle-hackle. Wing-ends are clear white from bars to tips, free from rust. Few cockerels are sound in this respect. Eyes should be red and legs willow.

Females are red in face and eye, with hackles silvery-white finely striped with black. Body and wings should be French grey lightly and finely pencilled with black, one even soft grey throughout, free from shaftiness. Breast salmon, shading off to ash-grey on thighs and underparts. Sound breast colour combined with good body colour is difficult. Many are too pale.

SILVER: For both male and female the colour is similar to Golden Duckwings except silver-white predominates in the male.

SEX-LINKED MATING

Duckwings were originally derived from Black-Reds and have continued to be so bred. It is quite usual to mate the two colours together. This provides sex-linkage, a Black-Red cockerel mated to Duckwing females producing Duckwing cockerels but Black-Red pullets. Don't use these pullets in breeding Black-Reds. Mark them carefully, and if they turn out to be sound in colour and free from dark caps mate them to a Duckwing cockerel to produce Duckwing pullets.

Two pens are advisable for breeding. To produce good cockerels, use a sound-coloured Black-Red male, perfect in black and with sound rich bay to end of wings, mated to Duckwing pullets as clear in neck hackles as possible.

Coarse markings and rusty wings will be an advantage for producing colour in males. Pullets from this pen would normally be **Black-Reds**.

For pullet-breeding, use a Duckwing male, pure Duck-wing bred, mated to sound-coloured Black-Red females. On sex-linkage theories this should produce, say, all "Silvers" (or Duckwings) in both sexes; but Duckwings are systematically crossed with Black-Reds, so purity is doubtful in both the silver of the Duckwing and the gold of the Black-Red. Therefore though you will produce good Duckwing pullets of show colouring, you will probably also produce some soft, even-coloured Black-Red females.

Duckwing pullets are also bred by mating sound exhibition Duckwing females to a medium-coloured Duckwing male.

Don't use in Black-Red pens any Black-Red females bred from Duckwings; so if you breed both colours toe-punch and ring all progeny, and be safe.

Self-Colours: Unfortunately the known self-colours (blacks, blues and whites) seem to have died out. They were excellent, and required no double-mating. Many years ago a special club catered for them, but the old hands died; and although one or two fanciers later became interested, staying power seemed lacking to create new strains.

They were all produced from crosses and sports in the first place, but all appear to have been pure Modern in blood. Though usually somewhat softer in feather (and perhaps longer in back) there were specimens seen that could hold their own in hot company.

Blacks were mainly produced from Brown-Reds and Birchens. Whites came first as olive-legged sports from Brown-Reds, and achieved their yellow legs from crossing with Piles. Blues were apparently created by crossing blacks and whites. There is ample reason for reviving them.

The standard for Blacks demands red faces and eyes, with dark even-coloured legs (not necessarily black), and jet black plumage with bright beetle-green sheen. Blues were permitted to have bright red or brown eyes, legs blue or dark olive, plumage light slate-blue with dark blue hackles, laced or unlaced as preferred. Whites had red eyes (no other colour permissible), snow-white plumage and orange or yellow legs.

Off-Colours: Blue-Reds and Lemon-Blues are occasionally seen, and are most attractive, but fail at present to entice new breeders—I don't know why. Lemon-Blues are like Brown-Reds but blue where the Brown-Red is black. Can you imagine anything more charming?

Serious defects include flat shins, crooked breast, duck feet, roach back and imperfect colouring. All merit the judge passing the exhibit.

Recommended weights: Male 20–22 oz. Female 16–18 oz.

These weights are a little higher than pre-war standards.

17

OLD ENGLISH GAME

These are the most popular of all bantams. They exist in multitudes of colours; and today, at all shows throughout Britain, large or small, no classes fill so well. This is not because (as so many pretend) they are miniature editions of, and descended from, our National breed of fighting game. It is doubtful whether the bantams have much large game blood in them—though it may well be that through the common barnyard bantam of the countryside they inherited, years ago, some game characteristics.

Figure 4. Spangled Old English Game Cockerel. Winner of Numerous Prizes for Mr. G. Edwards

Hardly a show fails to cater for these engaging little birds, and practically every colour in every breed of poultry is reproduced in them. Spangles are the greatest favourites, followed by Black-Reds, next by self-blacks and self-blues, Duckwings, Blue-Reds, Brown-Reds, Furnesses, Creles and numerous off-colours.

They are also bred muffed and tasselled (bearded and crested); and although many may consider these adornments foreign to miniatures of our national fighting Game fowl, it should be remembered that our bantams are really show birds, not fighters. In any case, muffs and tassels are present in the Large O.E.G.—the Bailey Muffs being famous fighters.

Breeding is interesting but not intricate. Multitudes of off-colours are produced without special matings, and double-mating for production of good cockerels and show pullets is not often practised.

Old English differ greatly from Moderns. Instead of height and reach they have cobbiness and normal low build. Broad strong fronts, well-developed shoulders, muscular thighs and active legs and feet are important.

LEGACY FROM FIGHTING DAYS*

The Old English Game Bantam Club requires the head to be small and tapering, with skin of face and throat flexible and loose. (This is a legacy from fighting standards. It prevents distress when hard-pressed in battle.) Beak should be big, the upper mandible shutting tightly over the lower, crooked or hawk-like and strong at base. Eyes are large, fiery, and fearless; comb, wattles and earlobes fine, thin and small (males are usually dubbed as explained elsewhere). The head is entirely different from the snaky style of Modern Game.

Neck should be strong and of medium length, with plentiful wiry hackles covering shoulders. **Back short, flat, broad at shoulders and tapering to tail** (somewhat heart-shaped); breast full and prominent, well-muscled, the breast-bone not deep or pointed. Belly small and tight, thighs short and muscular, following outline of body, slightly curved. Legs strong and close-scaled, with a good bend or angle at hock. Toes straight, with strong curved nails, the back toe extending backwards in a straight line. Spurs must be set low on leg.

Wings should be relatively large and powerful, and originally the tail was specified as "large", inclined to fan shape in the hen, and carried well up. A small tail tends to be the *norm* for the male O.E.G. bantam. Plumage hard, smooth and glossy, without much fluff.

Carriage should be defiant and active, ready for any emergency; agile and quick in movement, and in the hand well balanced, hard in flesh, with strong contraction of wings and thighs to the body.

Colours largely follow descriptions and methods detailed under Modern Game, but it should be noted that in O.E.G. *exact gradations* of colour are relatively unimportant.

---

* For a detailed consideration of O.E.G. standards see *Understanding Old English Game,* Batty J., The Spur Publications Company.

It is not possible to describe all sub-varieties or detail all matings. These have in any case been largely dealt with under Modern Game; but there is space to give particulars of a few popular colours.

Spangles: The popular Spangles are dark in colour, the foundation or ground colour being that of Partridge-bred Black-Reds. In fact, they are, to all intents and purposes, Partridge-bred Black-Reds with spangle-tips added. One pen will throw good specimens of both sexes, and they can be bred from a Black-Red cross as well as from Spangles on both sides.

Figure 5. Mr. J. C. Duckett's Champion Spangled Old English
Game Hen
Note the shortness of back, correct angle of legs and even spangling

All published standards require white tips to every feather if possible—the more spangles and the more evenly distributed the better; but a bird answering this description would do little winning nowadays. Spangling must now be small and sparse; and although as they get older, birds get gayer as spangles increase in size, they are required to be dark and not to show much

20

white. If, in fact, chicks come too gay it is usual to introduce darker colour by using Partridge hens.

Normal leg colour is white but occasionally yellow-shanked specimens are seen, and can win if of good quality. Eyes are red.

Black-Reds: These have two variations—dark (Partridge bred) and light or bright red (Wheaten bred). They are, of course, two different colour-varieties, and for show purposes must be treated as such.

Since the light or bright red colouring is preferred in cocks, more birds of Wheaten strain are bred than Partridges. Wheaten-bred cocks are bright orange in hackle, with saddle-hackle matching as nearly as possible. Wing bars and wing butts blue-black; back, saddle and wing bow bright crimson. Tail green-black, breast and thighs black. Secondaries to show rich chestnut bays to tips of wings. Legs willow, olive, white or yellow, but Wheatens should have white legs.

In Wheaten females the general top colour resembles wheat; and although this varies considerably, in general it is a soft cream colour, with the breast, underparts and thighs pale fawn. Neck hackle should be "clear red", tail well clipped together and nearly black. Primaries black or nearly so, fluff white, legs white.

The Partridge hen has golden hackle lightly striped with black, and reddish breast with ash-grey belly; primaries and tail dark, body-colour an even partridge shade lightly pencilled all over with delicate irregular black markings; the partridge colour to be of golden tinge rather than dark.

Partridge-bred cocks are much darker than Wheaten-bred, being dark red instead of bright red in hackles and much darker in back and saddle. The two varieties should not be mixed, but if Partridge-bred cocks are mated to Partridge hens, and Wheaten-bred cocks to Wheaten hens, they breed very true to colour.

Other Colours.—*Blue-Reds* or Blue-Duns are very charming. They may be bred pure or crossed with Black-Reds, or Brown Reds. Blue-tailed Wheaten females are among the loveliest of all bantams and are closely allied to Blue-Reds, which resemble Black-Reds except that blue is substituted for black. Faces and eyes are red, and white legs are preferred.

In blue-tailed Wheaten females the plumage is in all respects similar to normal Wheatens (see Black-Reds) except that tail, wing-primaries, etc., are blue, or shaded with blue, instead of black.

Blues are often bred with Blacks to retain depth of colour. Sometimes odd Black hens are simply put in the Blue breeding pen, or vice versa. Eyes may be dark or red, according to leg colour, which may be dark blue or white, and occasionally yellow.

Figure 6.   Black Old English Game Cock. Winner of Many Firsts
for Mr. A. R. Kerr

Blacks are of course mainly bred from Blacks on both sides.
Face and eyes may be red or dark. Black or white legs are usual,
yellow permissible.

Duckwings are bred pure, also from a Black-Red cross as
described in the chapter on Modern Game. This remark applies
also to Brown-Reds, Birchens and Piles, the breeding of which
is dealt with in the previous chapter.

Creles are a barred variety, sometimes bred pure but more
often by mating Partridge, Blue and Blue-Red hens to a Crele
cock. It is a delightful sub-variety which produces many variations (including cuckoo females) but which throws a goodly
proportion of mis-marked birds, not suitable for show though
useful for breeding.

Creles have red eyes; cocks have orange-coloured checkered
or barred necks, grey and white breasts; back and shoulders
orange, wing bows orange with dark-grey bars; secondaries bay
on outer web, primaries and tail dark grey; the plumage generally
checkered or barred. The hen's neck is lemon colour checkered

or barred with grey, breast and thighs checkered salmon, back and wings checkered blue-grey, with tail to match body colour. White legs are preferred and usual.

There are multitudes of variations, both off-colours and standardised. These would take too much space to describe fully.

So-called self-colours (blues and blacks, whites being seldom seen) may be pure colour or may have some white in sickles; but self-coloured cocks with red in saddle and hackle must be shown as brassy-backs—usually in A.O.C. classes. Furnesses are somewhat like brassy-backed Blacks but with more brassiness, and females are heavily marked with streaky greyish-fawn. They are not closely standardised, and once again type, build and carriage are much more important in O.E.G. than purity of colour.

Recommended weights: Male 22–26 oz.; Female 18–22 oz. These weights are slightly higher than old standards.

Serious defects include thin thighs, flat sides, crooked or pointed breast bone; duck feet, stork-type legs, in-knees; broken, soft, or rotten plumage; bad carriage or action, and indications of weak constitution.

INDIAN AND JUBILEE GAME
INDIAN GAME

As previously stated, in spite of their title Indian Game must not be shown as game, nor must they be entered in classes reserved for soft feather bantams. Unless therefore they have their own breed classification they can only be shown in sections reserved specifically for hard feather, or in the A.O.V. classes. They are the bulldogs of the fancy, and have been developed to a high state, having regard to the difficulties of reconciling bone and substance with miniature size in bantams.

The breed cannot be standardised as small as others, but serious attempts should be made further to reduce weights. Many specimens almost look like large breeds when staged in bantam pens. Shape and type must come first, then bone and substance, next colour and markings. Reduced size can be obtained by in-breeding and by special matings.

Good specimens can be bred in both sexes from one pen, but if standard lacing in females is really wanted (double-markings verging on triple-lacing) allied with almost self-black colour in males, the only way to achieve it would be by double-mating, as carried out in former years, but this would not be popular.

The **male** should have a broad head, beetle-browed, with short hackle just covering base of neck. A strong beak gives an appearance of power. The pea or triple type comb is small and close, the centre ridge rising above sides. Eyes are bold but not as cruel in expression as the Malay. Face is smooth and fine, the throat dotted with small feathers.

The very thick and compact body is extremely broad at shoulders, with very prominent shoulder-butts, not hollow-backed and not roached or rounded; the whole tapering to tail. and sloping downwards. Breast is prominent and rounded.

Wings are short and close, rounded at ends, carried high in front and closely tucked at tips. The medium-length tail has short narrow sickles and coverts, close and hard, carried somewhat drooping.

The very thick legs, with thighs round and stout, should have shanks which are short. Toes must not be twisted, and back toe must be well on the ground.

Figure 7.   Indian Game Cock. A Consistent Prizewinner for Mr. M. L. Goodfellow

This is an excellent specimen which shows compactness in a heavy muscular body

General appearance and carriage show power and solidity with activity and vigour. Carriage must be upright and commanding, the back sloping downwards towards tail; plumage very hard, close and short and the bird heavy and solid in hand.

Beak is yellow or horn striped with yellow, face a rich red throughout, eyes pale yellow to pale red. Pale eyes are usually preferred as giving a more typical expression. Legs are rich yellow or orange.

Body and neck generally are black with rich green gloss, neck and saddle hackles a little broken with rich bay or chestnut;

shoulders and wing bows beetle-green, with back and tail furnishings glossy green-black, all slightly touched or broken with bay or chestnut. Wing secondaries are bay on outer edge and glossy green-black on inner webs, primaries deep rich black with narrow chestnut outer edging.

The **female** resembles the cock in general characteristics. Tail should be short and carried low, though somewhat higher than in males.

Ground colour should be mahogany or nut-brown, the head, hackle and throat being glossy green-black, the under-hackle broken with chestnut. Primary wing feathers black, peppered in inner web with chestnut, secondaries black, outer webs toning with general ground colour.

The whole of the remaining body-feather should be very clearly laced with extremely glossy metallic green-black, the lacing distinct and double. An additional third black marking in centre of feather is often seen in the best specimens and is highly esteemed. The outer and inner lacing should be so glossy as to appear embossed.

Lacing on throat and breast should be particularly distinct, running off to indistinct markings under the vent. Back and tail coverts should be clearly marked, wing bows and shoulders being very distinct. Wing bars are usually peppered as well as laced.

DOUBLE MATING

The general method in cockerel breeding is to mate a cock of exhibition colouring with hens so heavily laced as to appear almost black rather than mahogany in ground colour. They need not be at all clearly laced—in fact if they approach self-coloured Blacks so much the better. These hens should have good sound wing bays, but be free from red or rust in hackles.

For pullet breeding the difficulty is to find the right cock—though after a few seasons of selective breeding any strain will produce them.

The process is to mate females of exhibition markings, with clear lacing, to a cock that in colour is the very antithesis of a show bird. If you mark and keep, in your pullet-bred youngsters, males that when young are laced in places like females, with red in hackle and general signs of pullet markings, you will be on the right tack.

In both matings of course you should select rigidly for build, substance and type; and in pullet-breeding particularly you will find it necessary to develop and build up your own strain. The cockerels that appear wasters in markings will often be your best pullet-breeders. It isn't wise to introduce outside blood, except with extreme care.

Weights recommended by the Indian Game Club are: Males 40–48 oz.: Females 32–40 oz.

These weights would be better reduced now that the bantams are well-established—about 4 ounces less now, with further reductions later, might be advisable, and could be attained while retaining shape and breed character.

Disqualifications are crooked backs, crooked beaks or legs, wry-tail and squirrel-tail, in-knees, red hackles, flat sides, walnut combs and single combs.

Other serious defects include crooked breasts, flat shins, rusty hackles, white in hackle, twisted hackle, long legs and thighs, heavy feathering.

JUBILEE INDIAN GAME

Jubilees should in every respect, except colour and markings, resemble Indian Game, and can be bred in the same pen.

Briefly, the male should be white with bay or chestnut markings. Head, underparts, neck, thighs, tail and coverts white. Breast white, but some chestnut lacing permissible. Shoulders and wing bows white, slightly broken with bay or chestnut in centres of feathers. Back white, touched on ends of feathers with bay or chestnut. Wing secondaries white on inner, and bay or chestnut on outer web; primaries white with narrow chestnut outer lacing.

Figure 8. Mr. M. L. Goodfellow's Jubilee Indian Game Hen. Her many wins include Championship of Bath and West Show

The hen should have a ground colour of rich chestnut or mahogany brown. Tail, head, hackle and throat white, pointed under-hackles white; double white lacing over the body where the Indian has black lacing, with gradations and variations of markings as described for Indians. Primaries are white, the inner web peppered with chestnut; secondaries white on inner web, outer web resembling ground colour but with a fine edging of white. Wing coverts are laced, but may be a little peppered.

White lacing should be as clear as possible and should look as if embossed or raised; but as the lacing has no gloss (as in the black lacing of Indian Game) this requirement is not often evident.

Weights, scale of points and defects set out for Indian Game may be taken to apply equally to Jubilees; and for breeding, the methods set out for Indians may be adopted. The two colours should be bred together, thus retaining a strong mahogany brown.

ASEELS AND MALAYS

Aseels, Malays and Sumatra Game (the last still surviving on the Continent) are the only other hard feathered breeds in bantams. The former are now seldom seen, and Malay bantams are very scarce. Breeders are almost entirely confined to Cornwall; and unless they can be further bantamised they are un-

Figure 9. Mr. Alex King's Prizewinning Malays, Male and Female.
(Photographer: Mark Stephens, Tunbridge Wells)
Note the three curves referred to in the text: this breed is worthy of a wider following

likely to regain former popularity. They are, of course, miniatures of a very tall, heavy breed; but bantams unable to stand upright and show their type and build in a pen 21 in. high do not commend themselves to many fanciers.

It is sufficient to say that Aseels and Malays together were largely progenitors of Indian Game, which originally were produced in Cornwall. They would in fact be better described as Cornish fowl, which is how they are named in U.S.A.

Aseels are pugnacious, pea-combed, and very similar to Indian Game. Bantam varieties formerly existed in Spangles, Reds and Whites.

Malays are chiefly distinguished for their extremely sparse, hard feather, cruel expression, overhanging brows, walnut comb, great length and substance of limb, and the typical three body curves—neck, back and tail forming a succession of three curves at approximately equal angles. They exist, in bantams, as Black-Reds, Piles, Spangles and Whites. Weights are 36-48 oz., the female being at the lighter end of the scale.

Chapter 4

SOFT FEATHER BREEDS

Bantams are mainly soft feathered, and may be divided into three groups—**Ornamental Varieties,** which include old-established true bantams such as Rosecombs, Sebrights and the purely "fancy" breeds; **Light Breeds,** which include all varieties of Mediterranean or similar origin, such as miniatures of large non-sitting breeds; and miniature **Heavy Breeds,** which cover all bantams of large utility type and heavy or general purpose character.

The Poultry Club classifies "ornamentals" as the following:

Belgian Bearded Bantams	Sebrights
Brahmas	Faverolles*
Cochins or Pekins	Houdans
Frizzles	Booted
Japanese	Creve-Coeur
Polands	Yokohama
	Rosecombs

The Booted, Creve-Coeur and Yokohama are rare varieties of bantams being seldom seen. The remainder are described in this chapter. What should be appreciated is that there is a mixing together of "heavy" and "light" breeds in the ornamental classification.

ORNAMENTAL VARIETIES

Some of these varieties (such as Brahmas and Cochins or Pekins) might be classed as miniatures of Light or Heavy breeds but are included in this section because of their markedly decorative or ornamental character.

Exact segregation into sub-sections is not easy. Hamburghs for instance might be included here instead of with Light breeds of utility type. However, they are now accepted at weights that really remove them from genuine fancy bantams and make them more suited to the miniature Light breed section.

Many ornamental varieties (such as Rosecombs, Sebrights, and Japanese) are true bantams genetically. These have no

* Now reclassified as a heavy breed, not an ornamental.

counterparts in large breeds, and were probably evolved entirely from pure bantam blood in the first place. That is why they show pronounced genuine bantam characteristics, such as the long, low, drooping wings that are distinctly out of place in more recently bantamised editions of large breeds.

BELGIAN BEARDED BANTAMS

There are only two varieties of Bearded Belgians standard-ised in Britain, but what a mulitude of variations they present in colour and markings! They are old-established true bantams, without counterparts in large breeds. Each of the two main varieties (Barbu d'Anvers and Barbu d'Uccle) has many colour-varieties; some intricate, and all extremely attractive.

There are also several other varieties of Belgian Bantams (bearded and otherwise) not standardised in this country and therefore not described here.

The descriptions given are translated from the standards of the Belgian National Club. These have in the main been adopted by the British Belgian Bantam Club.

Barbu d'Anvers: The d'Anvers is always rose-combed and clean-legged, with rather large head and short, curved beak coloured in keeping with plumage. The broad-fronted comb is curved, ending in a spike or leader at rear; preferably covered with fine work or points, but alternatively it may be hollowed and ridged. The leader follows the line of the neck.

Figure 10.   Mr. E. C. Ellis's Barbu d'Anvers Male. A Noted Prizewinner

Eyes are dark and vary in colour with plumage. The brow is heavily feathered, and the face covered with feathers which stand out, slope backwards and form whiskers covering ears and earlobes. (Neither Continental nor British standards are clearly explanatory on this point. The standard description of facefeather in d'Anvers suggests that it forms part of the beard and whiskers. See also note regarding feather on brows in d'Uccle standard.)

The beard is composed of feathers turned back horizontally from both sides of the beak, and vertically downwards below the beak; the whole beard and whiskers forming a collar and muffling the face. (*Note*—in the Belgian National standard the beard is required to form a trilobe, as in the Barbu d'Uccle).

Earlobes are small and wattles either rudimentary or absent —the latter preferred. Thick hackles, convexly arched, entirely cover the back and base of a neck moderate in length, and form the typical "boule" or bull-neck, almost joining in a close cape at front.

The body is short and broad, with prominent breast carried high; the back short and sloping down to tail. Wings are medium in length, sloping downwards; tail almost perpendicular, main feathers strong, sickles narrow, sword-shaped and only slightly curved.

Thighs are short, shanks medium in length and free from feathers. The toe-nails match beak in colour.

In appearance the male is small, proud and upright, the head thrown well back as if ready to crow.

General characteristics of the hen resemble the cock; but the head appears broader and more owl-like. The hackle also forms a ruffle behind the neck, with broader feathers; and hackle, contrary to that of the cock, diminishes in thickness towards base of neck. The tail is short, carried slightly upwards; and the hen's appearance is compact, plump, small and lively.

There are no particular problems in mating, and in most colours good birds of both sexes can be bred from one pen— though some of the colours would probably be improved by a definite course of double-mating. Care should, however, always be taken to breed for full beards, muffs and whiskers, and to preserve the typical body-style by maintaining short backs and gaily-carried, well-developed tails. The desired neck-hackle also can readily be lost as a breed characteristic if not carefully preserved.

The British Belgian Bantam Club does not publish or advocate an actual weight standard, but purely as a general guide suggests—male 24–28 oz.; female 20–24 oz.

Serious defects include strongly-developed wattles, conspicuous earlobes, wry tail, squirrel tail and excessive length of leg.

Disqualifications: Wattles cut or removed, single comb, absence of beard or whiskers, feathered shanks or feet, more than four toes, yellow colour in legs, feet or skin, and any form of faking.

Barbu d'Uccle: The d'Uccle is always single-combed and feather legged. The cock's head is small, beak short and curved; comb single, small, upright and fine, evenly serrated, blade following line of neck. Eyes surrounded by bare skin. Brow heavily covered with feathers lengthening towards rear and tending to join behind neck. (In both Continental and British standards this description of brow feather in d'Uccle is missing in d'Anvers, where the face feather, not brow feather, is required to show a tendency to join behind the neck. Greater clarity might be possible on several points in the official standards of the two bodies.)

The beard is full, with long feathers turning outwards, backwards, and downwards and forming a trilobe (three ovals in a triangular group). Earlobes should be inconspicuous, wattles as small as possible.

The silky neck hackle has a tendency to form a thick mane behind the neck, covering shoulders, saddle and back. The body is deep, with very broad breast carried high and forward. Wings fit tight and close, sloping downwards to abdomen, but not beyond. Ends of flights are covered by the abundant, long saddle hackle, and the well-furnished tail is carried almost perpendicularly, the main sickles slightly curved.

Figure 11.   Cuckoo Barbu d'Anvers Female. Bred by Mr. E. C. Ellis

Legs are short and well apart, hocks covered with long stiff feathers inclined downwards. Front and outside of shanks must be furnished with feathers lengthening towards footings; the foot feather horizontal, curved slightly back, outer and middle toes covered to ends.

Carriage and appearance—a little cock with majestic manner, short and broad, with heavy plumage development.

The hen resembles the cock except for usual sex differences, but the beard is formed with softer and more open feathers. Neck-hackles are very thick and composed of broader feathers, the tail short and not carried high, the lower main feathers diminishing evenly in length. General appearance short and cobby.

There are no particular mating problems. They are seldom double-mated, though there is no doubt some of the colours and breed characteristics could be greatly improved by a short intensive course of breeding from two pens.

As with the d'Anvers the British Belgian Bantam Club does not publish a weight standard, but the following guide is suggested—males 28–32 oz.; females 24–28 oz. There should be usual variations for age and maturity.

Serious defects include strongly-developed wattles, conspicuous earlobes, wry-tail, squirrel-tail, long legs.

Disqualifications are—wattles cut or removed, comb other than single; absence of beard or whiskers; poorly-feathered shanks or feet, more than four toes, yellow legs, feet or skin, and any trace of faking.

Most but not all of the colours listed apply to both d'Anvers and d'Uccle, so there is no need to repeat them for each, or to distinguish between them. Readers will realise that references to foot-feather apply only to d'Uccle, and so forth.

Belgians exist in multitudes of colours and colour-patterns— far more than in any other breed. Main colours only are described and for particulars of others mentioned but not fully detailed reference should be made to the illustrated *British Poultry Standards.*

In a number of items official standards are either not clear or are somewhat contradictory. This sometimes happens when attempts are made to describe points too fully—clarity is often lost.

In all colour-varieties comb, face, earlobes and rudimentary wattles are red. Legs and feet vary with plumage colour, from black or slate-blue in dark colours to white in Cuckoos and Whites.

The Millefleur cock shows very attractive and intricate markings and colour. The head is orange-red with white points, the beard black laced with light chamois, each feather finishing with a white triangular tip on a round black spot. The neck feather is black, broadly fringed with orange-red and with gold shafts, each feather having a black end finished with a white point. Great abundance in neck hackle makes the neck appear orange-red, the black parts being scarcely visible.

The back is red, shading off to orange at saddle. Wing-bows are mahogany-red, each feather tipped with white; wing-bars russet-red, with regular cross-bars formed of lustrous green-black pea-shaped spots finished with silver-white tips. Primaries are black with a thin outer edging of chamois; the visible lower part of each secondary is chamois, the upper part black. Remainder of wing is chamois, each feather having a large pea-shaped white spot on a black triangle, evenly spaced to conform to wing-shape. (Notice the reversal of these pattern-markings from the normal—though the distinctions are seldom complete or clear throughout.)

Tail is black with metallic-green lustre, fine edgings of dark chamois and white triangle-tips. Breast and remainder of plumage (including footings in d'Uccle) golden-chamois with light chamois shafts, each feather tipped with a white triangle on a round black spot. Beak and nails are slate-blue, eyes orange-red with black pupils.

The Millefleur hen has black pea-shaped spots tipped with white triangles on golden-chamois ground colour. Tail feathers are black, finely laced with chamois and with white tips. Wing markings and other plumage are as described for cock, allowing for sex variations.

Figure 12.   Pair of Millefleur Barbu d'Uccle. Consistent Prizewinners for C. and V. Mayhew

This splendid pair show quite clearly the "boule" required in this breed

Defects to be avoided in both sexes are washed-out or light ground-colour and gay or uneven markings. (Gay markings show too much white.)

The Porcelaine is an extremely delicate colour-pattern. Markings are as described for Millefleurs, substituting a light straw for the chamois ground colour; and the pea-shaped spots are a very pale blue instead of black. This pale blue is substituted for black throughout, and is the true-breeding form of blue mentioned elsewhere in this book—not the Andalusian type of diffused black and white.

Eyes orange-red with black pupils. Beak and claws slate-blue.

Very light or washed-out ground colour should be avoided, also gay or uneven markings.

The Quail cock is very striking, with head feathers dark green-black finely laced with gold, beard golden-buff or nankin, shading darker towards the eyes, where plumage is black laced with gold. Hackles on neck are brilliant black, sharply laced with lustrous golden-buff and with pale-buff shafts.

The back is black, with gold lacing starting at middle of feathers and narrowing towards tips, forming lance-points of gold with well-defined shafts of light ochre from root to point; feathers relatively broad under the neck hackle but longer and narrower towards saddle, where colour and black ground are more intense.

Wing-bows are light gold, the lower half of each feather black, upper half nankin or buff. Wing-bars are light ochre, each feather with black triangle-tip, the triangles forming two regular bars. Secondaries have chamois lower web, black upper; and primaries are dull black.

The tail is black with metallic green lustre, finely edged with brown and with faintly-defined light shafts. Sickles are black, furnishings black laced with chamois and well-defined pale shafts.

Breast colour is important. Each feather is nankin finely laced with yellowish-buff or ochre, shafts being distinct and clear. A breast clear of black is very desirable, and the colour should shade off at thighs, abdomen and underparts to greyish-brown with silky golden tips.

The Quail hen has ground colour of umber on head and neck, with very fine gold lacing; the neck velvety, darker than the back and clearly distinguishable from it, shafts and lacing clearer and more golden towards breast and back. Back covered with dark umber-coloured feathers with silvery-velvet lustre, finely laced with chamois and with strongly-contrasting nankin shafts.

Wings are dark umber laced with chamois, brightening in colour towards ends; primaries dark umber, tail and cushion similar to back.

35

The breast is nankin, with pale, distinct shafts, finely and progressively bordered towards sides with dark umber forming a laced colour-pattern.

Eyes in both sexes are dark brown or nearly black, with black pupils; legs and feet slate-grey, beak and nails horn-coloured.

The general effect in Quails is that upper parts are dark and lower parts light, giving the appearance of a dark cover or cloak. The dominating dark tint is umber, and the general light tone nankin or yellow ochre. Well-defined pale shafts are important.

The chief colour-defect to avoid is any trace of salmon or brownish colour on breast.

The Blue Quail is similar to the Quail in most respects, but umber colour and markings are replaced by blue.

The Cuckoo is uniformly cuckoo-coloured, with transverse bars of dark bluish-grey on a light grey ground, each feather having at least three bars.

Eyes are orange-red; legs, feet, beak and toes white, often mottled with bluish-grey in young birds.

Defects to be avoided are white feathers, feathers spotted with white, all-black feathers, red on wings, shoulders and hackle.

Black mottles have all feathers black with metallic green lustre, evenly tipped with white, tips varying in size with plumage. Avoid gay markings or uneven distribution.

Eyes are dark red, legs and feet slate-blue or nearly black, beak and nails dark horn.

The Black has sound-coloured plumage all over, with metallic-green sheen, and without false colours. Eyes are black, legs and feet blue or nearly black, beak and nails black or very dark horn.

The White should have pure white plumage throughout, without straw or yellow tinge on back, and without false colours. Eyes are orange-red; beak and nails white.

Other colours include Laced Blues (Andalusian type—a diffusion of black and white); Self-blues (true-breeding pale blues evolved from the blue markings of the Porcelaine) and Blue-mottles (similarly marked to Black-mottles). Other colours, though not available at present, are Ermines (black-pointed whites); Fawn-ermines (black-pointed buffs), Partridges, Silvers, Golds and Spangles.

There are also other Belgian bearded bantams not yet standardised here, the most notable being the Watermael, which has a small "flying" crest or tassel, a tri-lobed beard, and neat rosecomb with three distinct and separate spikes or leaders following the neck. The Watermael is clean-legged.

BRAHMAS

Brahmas are miniatures of the large "Brahmapootra" breed once known as "Lords of creation", and were formerly extremely popular. Like Pekins they are Asiatic in character, and have several closely-related characteristics, though less fluffy and not so plentiful in feather. Unlike Pekins, however, very low carriage is not desired. The male has a typical up-standing style, but females should be lower in build.

We have only two colours—Lights and Darks. Lights are black-pointed or "Columbian" in markings; Darks are a study in black and silver, the hen's pencilling on a soft grey ground colour being particularly attractive. Because of this pencilling Darks would be better double-mated, but Lights will produce good birds in both sexes from one pen.

They are good layers, good mothers, hardy and easy to rear. Both varieties have triple or pea-combs and small skulls with rather prominent "Asiatic" eyebrows. Beak is short and strong, eyes large and prominent, face, earlobes and wattles fine, smooth and free from hairs or feathers.

Cocks have long necks covered with flowing hackles. There is a typical Brahma depression at junction of back of head and upper hackle.

The body is broad and deep, with full breast and level keel; short flat back with saddle rising to the tail, which is medium length and carried nearly upright; main feathers well spread, coverts curved, abundant and broad, hiding the quill feathers.

Legs are moderately long in males, shorter in females, well apart and feathered. Thighs are large and hidden in front by lower breast-fluff, the thigh fluff abundant and standing out. Hocks are covered with soft feathers, which may be harder if accompanied by heavy shank feather and footings. Vulture hocks are not desirable but are not a disqualification. Profuse shank feather stands out from legs and toes to ends of middle and outer toes. Carriage is upright and active, of majestic manner.

The female is similar generally (allowing for differences in sex characters) but lower-built and shorter in legs and neck.

COLOUR STANDARDS

Colour and markings are very important, being given 40 points out of the 100.

The Dark cock is silver-white in head, neck and saddle-hackles, with a sharp stripe of brilliant black to each feather, free from white shafts and not running through at tips. Breast, underparts, thighs and fluff intense glossy black; back silver-white, but glossy-black laced with white between shoulders. Wing bows silver-white, primaries black with some white in outer edges, secondaries with white outer webs forming wing-bays. Coverts glossy black, showing a very distinct wing-bar. Tail black, tail-coverts edged with white. Leg-feather and footings black, but white in small amounts is permissible.

The Dark female has silver-white head, with neck hackles as described for cock, or pencilled in centres. Tail black, edged with grey or pencilled. Remainder of plumage clear soft grey, pencilled with concentric rings of black or dark-grey markings following outline of feather; pencilling sharp and uniform, and three rings or more if possible on each feather.

Figure 13. Male and Female Dark Brahma Bantams owned by D. W. Ledward

The Light cock should have head and neck hackle as described for the Dark variety. Saddle white, but may be slightly striped with black if hackles are very dense. Primaries black edged with white, and secondaries black on inner web and white on outer. Tail black, or edged with white. Remainder of plumage clear white, with some black usual and permitted in footings. Under-colour may be white, blue or slate.

The Light hen has silver-white neck-hackle densely striped with black, the black centres entirely surrounded by a white fringe. In other particulars the female resembles the male.

When breeding for perfect markings in the Dark variety, it is wise to use methods adapted from those described else-where for partridge-coloured birds. The great difficulty is to produce sound black breasts and solidly-striped necks in males, with fine pencilling in females. The problem is not so severe as in Partridge Wyandottes, partly because black striping is general in the neck hackle of Brahma females; but as a general guide, to breed good females a cockerel with ticked or laced breast would be mated to well-pencilled hens, and for cockerel-breeding a male with soundly striped neck and solid black breast would be mated to females with striped necks and dark colour, having heavier black markings instead of fine concentric pencilling.

Other details of mating would follow the methods in Partridges, or would be gathered by experience in building up your own strain. It is impossible to describe them fully here.

The Poultry Club recommended weights are—male 28–32 oz.; female 24–28 oz.

Serious defects: Comb other than pea-type; twisted hackle; faulty wing feather; lack of leg feather and footings; red, yellow or buff in plumage.

COCHINS OR PEKINS

The Cochin or Pekin bantam was imported from Pekin in China, about the middle of last century. It existed under the name "Pekin" until comparatively recent years, when, largely owing to American influence, the Club catering for it adopted the duplex name; the obvious intention being to model the breed on the large Cochin, which had in its turn once been known as the Cochin China fowl.

When first imported buff was the only colour established. Nowadays we have Buffs, Blacks, Whites, Partridges, Blues and Cuckoos. Silver-Pencils (a lovely variety) were also once produced but were allowed to die.

The original Pekins were obviously not at all like present-day birds. They lacked the modern plentiful feather, were narrow and high, long on leg and long in back, with vulture hocks and much bigger size.

The Pekin bantam nowadays can claim considerable superiority over the large breed. Whereas big Cochins have gone up on leg and achieved a height too great in proportion to length of body, the bantam is short in shank and low to ground, with much more pretension to globular shape, especially in females.

The most popular colour is the Black, though when bred in good feather there is a charm about Buffs that is missing in other varieties; and probably Whites have considerably more feathers per square inch than either.

The trouble necessary to keep them pure in colour, and the washing required for show, prevents Whites becoming the most popular; and certainly when we see the filthy grey colour acquired from partitions of dirty metal show pens (even at some of our classic shows) we can understand why they are not greater favourites. Yet a well-washed, rounded, matronly white hen is the most pleasing exhibit of all.

Partridges, alas (lovely in colour as they are), are almost extinct; but lately Mottles are gaining ground and may replace the once-popular Cuckoos.

VARIETY OF CONDITIONS

You can keep this breed under almost any conditions. They thrive on grass runs, lawns, or sand, and intensive conditions suit them more than most breeds. Their foot-feather is not a drawback, though when maintained in show form they need careful penning and lightweight litter. Sand and sawdust are excellent floor-coverings.

Figure 14. Pair of Black Pekins. Bred by Mr. E. C. Ellis
A beautiful pair of prizewinners which are typical of the breed

When in the breeding pen, to ensure fertile eggs it is customary (and indeed necessary) to trim plumage strongly. Tails and vent-fluff are heavily cut in females, and in males the fluff *below* the vent is cut away and footings trimmed short to enable birds actively to mate. This means that for summer shows exhibition birds are quite distinct from breeding stock.

Pekins are excellent layers of eggs that are usually tinted and very large in comparison to size of bird. As they are often kept semi-intensively in winter, they lay early in the year, and early-hatched, healthy chicks are thus general. In the bad old days they were often hatched very late, to achieve small size for show, but nowadays we realise that late hatching perpetuates weedy stock.

They are excellent broodies in spite of their feathered legs; and there is probably no other bantam that breeds so true if the initial stock is sound. They are extremely docile, and their usual precocity in babyhood has a great advantage, because they are catered for by a live club which encourages many classes for baby chicks at summer shows.

The charm of baby Pekins shown at ages from 12 to 16 weeks is almost unbelievable. True, many of these precocious youngsters develop too fast to make good adult show stock; but they can be excellent breeders, and in any event have served their turn in providing pleasure and competitive interest during the summer show season.

The head of a good Pekin cock is small, beak short and slightly curved. The single comb (often too large in show specimens, particularly when penned for extensive periods) should be firm, erect and evenly serrated, with an even top curve. Face, wattles and earlobes should be smooth and fine, the lobes nearly as long as the wattles, which should be long and ample but are sometimes too circular in shape.

The neck is short and carried well forward, hackle long and abundant. The short, broad body is globular or "ball shaped", with deep full breast, the back increasing in width to a full saddle rising well from the shoulders and plentifully furnished with long soft feathers. Wings are small and tightly tucked, with ends hidden by saddle hackle. (Very frequent faults are badly-slipped or split wings.)

Tail is full, soft and without stiff feathers, with very plentiful coverts hiding the main quill-feather, the whole forming one unbroken cushion curving from back and saddle.

Thighs are stout and hidden by a ball-shaped mass of fluff. Legs are short and well apart, covered at hocks with soft feathers curling round the joints. The short thick shanks are well furnished with soft feathers standing well out, and the feet are plentifully covered with feather to ends of middle and outer toes. In good specimens the inner toe is also often feathered.

Carriage is very important—it should be low-fronted, with head almost at a level with the tail.

The female is similar to the male except for usual sex distinctions, but the back rises to a fuller and more rounded cushion.

A point in which good birds in both sexes excel is in lower frontal feather, the plentiful soft plumage of the lower breast hiding the legs and merging with footings to give an almost shankless appearance. There must be no suggestion of "cup

41

and saucer" outline, such as that caused by truncation or segregation between body and footings.

Beak, legs and feet are yellow in all colours. In Blacks and other dark colours they may be shaded, but willow or green legs are not permissible. Eyes generally red or orange, but red preferred. Comb, face, earlobes and wattles bright red.

In *Blacks* plumage is rich black throughout, with lustrous beetle-green sheen, free of white or coloured feathers. (The Club does not penalise light or white undercolour in males, so long as it does not show through. This is to avoid the necessity for double-mating common to black breeds with yellow legs.)

Whites should be pure white, free from cream, yellow tinge, ashiness or black splashes.

Buffs should have sound buff plumage in both sexes, even in shade, sound to roots of feathers and free from bronze, black or white. Exact shade of buff is not material so long as it is free from shaftiness, but paler shades are preferred.

Cuckoos: Plumage light French grey, evenly and distinctly barred across each feather several times with dark slate. Markings regular and fine, free from white, straw tinge or brassiness.

Partridges: Head of male dark orange-red. Neck hackle bright orange or golden, paling towards shoulders, each feather striped down centre with black. Back, shoulders and wing bows rich crimson. Saddle hackles to match neck as nearly as possible. (A point of excellence is for hackles to shade off to pale lemon colouring.) Striping should not run through, and hackles should be free from black fringing or tipping.

Breast, thighs, underparts, wing butts, tail and coverts, foot and shank feather are lustrous green-black. Primaries and secondaries have bay outer and black inner webs, wing ends showing a sound triangle of bay colour when closed.

Head and neck hackle in the hen are light gold or straw, each feather striped down the middle with black. Remainder of plumage light golden brown, finely and evenly pencilled all over with three or more concentric rings of glossy green-black, the whole of one even shade and markings.

Mottles: These were formerly called Black Spangles, and resemble Blacks except that each feather is tipped with a small white spot or spangle, the whole as evenly mottled as possible: the amount of white showing to be small and not nearly equal to the black ground colour. Wings should be similarly marked but are usually more or less white-splashed.

Double mating in Pekins is not necessary except in Partridges. (It is not possible to produce exhibition birds in any "partridge" variety in both sexes without using two pens. For this see methods described elsewhere.)

Sometimes, however, in Blacks a modified form of double mating is used, to maintain sound undercolour in males and to produce very soft, broad plumage in females; but speaking generally excellent results can be achieved by mating a good cock, excelling in green sheen, with two kinds of hens—one kind with plentiful green sheen and rich colour, the other kind duller but as broad and soft as possible in feather, particularly on saddle and cushion.

In Buffs do not breed from males with three distinct shades, and do not mate birds contrasting badly with one another. Avoid lacing if possible, and don't use males with very dark cinnamon in tails and wings. Only experience with your own strain will tell you just how to mate, but speaking generally the male will be darker than females.

It is also more important to avoid white in plumage than to avoid traces of black; and if for breeding you can select birds without white quills in flights so much the better. Don't confuse pale buff undercolour, however, with white. A bird can have very pale underfluff and still be pure buff.

Blues are single-mated and may be any shade of rich blue, but pigeon blue is preferred in females, with darker hackles in cocks. The Club Standard states they should be clear of lacing, but attractive laced birds are sometimes seen.

In mating Pekins generally, whatever their colour, select for short back, quick rise of saddle and tail, broad saddle and extremely plentiful feathering on shanks, avoiding birds short of middle-toe feather; and don't forget the globular shape.

Contrary to general belief, plentiful feather on feet and shanks is much more desirable than extreme length, which is not greatly esteemed. Owing to its feathered shanks this breed is subject to scaly legs.

Approved Club weights not to exceed: Males 24 oz.; Females 20 oz. Exact weights are not very material in practice.

Serious defects include twisted or drooping comb; legs other than yellow (except for dark shading in dark varieties); slipped or badly-split wings; eyes other than red, orange or deep yellow; any deformity.

FRIZZLES

Frizzling as a plumage character can be introduced to any breed by a simple cross, but the Frizzle is a distinct breed. Speculation as to their origin tells us little. In spite of their being standardised here as a breed they have never become popular.

It is often imagined that curling irons are used in preparing them for show, but this is quite unnecessary—it is easy to breed birds with plenty of curl. After ordinary washing, stand them facing the fire and see the plumage rapidly resume its natural curl.

There are no problems in mating, except that prolonged breeding from frizzled plumage on both sides produces weak feather. It is therefore customary to use in the breeding pen a proportion of the plain-feathered birds produced by ordinary matings. Some of these should always be retained for breeding.

43

Figure 15. Champion White Frizzle Bantam Cockerel owned by
Mr. and Mrs. W. B. Johnson

They are excellent layers and good broodies, and require
no special care in housing and maintenance.

The male has a fine skull, short strong beak, single upright
comb of moderate size, smooth face with medium-sized wattles
and lobes, and full bright eyes.

The neck is moderately long with abundantly-curled
hackles showing a decided "mane". The body is short and
broad, with round and full breast, long and drooping wings.
Tail is full and loose, carried high, with full sickles and plentiful
side-hangers. (Note: Lyre-shaped sickles were once considered
correct in cocks but are not standardised.)

Legs are fairly short, set well apart, with shanks free from
feathers and four fine, well-spaced toes. Carriage is erect, active
and strutting.

Plumage should be crisp and long, each feather broad and
curled backwards towards the head; the curl close, abundant
and even, with hackles full and well-curled.

Male and female are alike in general characteristics, but the
female comb is smaller in proportion, and the "mane" is smaller
due to less neck hackle.

Colour: Beak yellow or horn; legs and feet yellow in Buffs,
Reds, Columbians, Whites, Black-Reds, Mottles, Duckwings
and Cuckoos, and black, blue or willow in Blacks, Birchens or
other dark varieties. (There are variations in leg colour, and

yellow legs are sometimes insisted on in Blacks, though not standardised.) White legs and beaks are permissible where so standardised in O.E.G. colours. Eyes should be brilliant red. Comb, face, wattles and earlobes bright red.

Plumage pure and even throughout in self-colour varieties. Columbians as in Wyandottes, and other "natural" colours as in O.E.G.

Suggested maximum weights: Male 24–28 oz.; Female 20–24 oz.

Serious defects include want of curl, comb other than single, drooping comb, white lobes, narrow or soft feather, long tail, or any deformity.

JAPANESE

Japanese have no counterpart in large breeds, and are the shortest-legged of all poultry, shanks also being so tiny in good specimens that breast-bones almost rub the ground when walking. Their waddling gait and peculiar dwarfish character are allied to the charm of obvious antiquity. They are very small, and there seems no limit to the tiny size that can be achieved; one case being authenticated of a hen 8 ounces weight laying eggs of $1\frac{1}{2}$ ounces.

They have the shortest of backs, their vertical tails and backward carriage of head giving them a "hairpin" build. They have very large combs (especially in males) which accentuate their dwarfish character. Their combs are, indeed, almost freakish in size compared with the birds themselves.

In spite of their unusual build the novice need not fear tackling them, so long as they can be provided with protection against wet grass and inclement weather conditions. They are long-lived, and cocks are often at their best about their fourth year; in which they differ from many one-season show birds. Good specimens can be bred in both sexes from one mating, and after the first couple of months or so chicks are really hardy.

Chief points are shape, carriage, feather and low build. A bird seen from the front should show feet only, no shanks being visible. Essential features are short legs, good tail carriage, short back and large, broad head. These account for 55 points in the Club standards.

In general characteristics they should be very small low-built, broad and cobby, with deep full breast and full-feathered upright tail. Appearance is somewhat quaint, due to very big combs and waddling gait. Plumage is very full and abundant.

The head is large and broad, beak yellow, short, strong and slightly curved; comb single, the larger the better, broad based, thick, fleshy and coarse in grain, with four or five broad-based spikes. Lobes and face smooth and red, face full over the eyes. Wattles large and pendant, eyes fiery orange or dark red.

Figure 16.   Black Tailed White Japanese Male: A Club Champion.
Bred by Mr. W. J. Kemp

The Standard requires a short and very full-feathered neck, with plentiful hackles covering the shoulders, and a back very short and broad, shaped like a narrow U from the side, tail and neck forming the vertical sides, the hairpin shape thus given being nearly lost in full-feathered males.

The round, full breast is carried rather raised, and the very large and upright tail has main feathers rising above the head by about one-third of their length and well spread. Main sickles, long and sword-shaped, are only slightly curved, with numerous soft side hangers.

The tail may just touch the back of the comb, but the bird must not be squirrel-tailed. Saddle-hackles should be plentiful and long; wings low, with tips of secondaries touching the ground immediately under stern.

Legs are very thick and short, smooth, yellow, free from feathers and hardly seen. Thighs are short and not visible.

The female and the male follow the same general description. The hen's comb should be large and well serrated, and if so large as to droop to one side this is a good point. The well-spread tail reaches well above head level, with the foremost pair of main tail feathers slightly curved.

Main colours are Black-tailed Whites, Mottles, Buffs, Blacks, Blues, Whites, Greys, Duckwings and Black-Reds, but there are often numerous off-colours.

46

Black-tailed Whites have wing primaries and secondaries black edged with white, the wings showing white only when closed. Tail black, sickles and coverts edged with white. Remainder of plumage white. (Sappiness, black in hackles or white splashes in tail are defects.)

Mottles are black with clean white tips to feathers. Lacing on plumage is a defect.

Figure 17. Mr. W. J. Kemp's Black Tailed White Japanese Pullet. Winner of Many Prizes

Buffs are buff-columbians, similarly marked to black-tailed whites, otherwise are even shade of buff all over. Self-coloured buffs are unknown and are not standard.

Greys are a very attractive mixture of black and silver, the cock's breast and hen's plumage laced with silver.

Blacks, Blues, and Cuckoos follow usual colour-standards, and most of the "natural" colours agree with Game Standards.

Single-mating is practised, and it is of advantage to the beginner that double-mating is not necessary—though probably the more intricate colours might be improved by it.

Suggested maximum weights: Male 20–24 oz.; Female 16–20 oz.

Weights much lower than these are often known.

Serious defects are white in lobes, narrow build, long legs, long back; wry-tail, squirrel-tail, tail carried low; deformed comb, lopped comb in males; high wing carriage, soft sickles, shanks other than yellow or shaded yellow.

Special note.—In Europe Japanese bantams are also standardised with both frizzled and Silkie types of plumage.

47

POLANDS OR POLISH

These are miniatures of a very old crested breed, supposedly of European derivation—though it is not certain whether the name refers to their crests or their alleged country of origin.

They exist in Chamois (white-laced buff), Golds, Silvers (each laced with black), Whites, and White-crested Blacks. A few White-crested Blues are also seen.

The great feature of the breed is the crest, which falls evenly round the head in males (and should be large) and is globular in females. They are much more hardy than might be imagined, and are usually very long-lived, good layers and breeders. No double-mating problems are involved, so they are a good breed for novices who fancy ornamental birds.

A good bird has a large head with a pronounced protuberance on top of the skull. From this springs a full crest, circular on top, free from split or parting, compact in centre and falling evenly round the head with untwisted feathers. (In the hen the crest must be globular.) Beak medium, nostrils large and very prominent, rising well above curved line of mandible.

Figure 18.   Mr. and Mrs. W. B. Johnson's White Poland Hen.
Winner of many First Prizes

Comb should be horn-type and very small. (Absence of comb is an advantage.) Eyes are large and full, face smooth and beardless in white-crested colour-varieties, completely covered by muffling in other colours. Earlobes should be round and small, not visible in muffed varieties. Wattles in white-crested colours to be of fair size, but absent or mere rudimentary in other varieties.

The neck is long, with plentiful flowing hackle. Body full, round and fairly long, shoulders wide, back flat and tapering. Wings are large and closely-carried, sickles and coverts plentiful and well-curved in the cock, tail full and neatly spread, carried somewhat low.

SPRIGHTLY CARRIAGE

Legs are fairly long and slender, shanks free from feathers, with four slender toes well spread. Carriage is erect and sprightly; and except whether otherwise stated, general characteristics of male and female are similar, allowing for sex variations.

A beak dark blue or horn is obligatory in all varieties. Comb and face red, eyes red, earlobes blue-white; legs and feet dark blue.

Figure 19. Charming Family Group of White Crested Blue Poland Bantams. Consistent Winners for Mr. A. Stafford

The Chamois is clear buff ground with white lacing, the cock's crest white at root and tips, but free from wholly white feathers—though in adult and ageing birds these are not heavily penalised. Muffling mottled or laced, not solid buff. Hackles may be tipped with white, but tail sickles, coverts, wing bars and secondaries should be laced. The hen should as far as possible be white-laced on a buff ground all over, including crest.

49

The Gold is clear golden-bay with black markings, with crest black at roots and tips, and free from wholly white feathers. Hackle and primaries are tipped with black, remaining plumage laced with black. The hen should be golden bay, laced with black all over.

The Silver is as described for Golds but with silver-white ground colour.

The White should be pure-coloured throughout.

The White-crested Black should have rich metallic-green sheen on sound black plumage, with snow-white crest. White-crested Blues are similar, but the even-shaded plumage may be either laced or clear.

It is popularly but wrongly believed that White-crested Blacks and Blues should have clear white crests, as erroneously described in former standards. This is not so. In these there should be a broad band of colour above the beak and at base of crest, composed of feathers the same colour as body-plumage. Absence of this coloured band at lower front of crest means that birds have been plucked, and should be disqualified.

It should be repeated that white in crests of laced varieties is usual in birds over one year old, and should not be penalised; but wholly-white feathers are not desirable.

There are no particular difficulties in breeding, except that stock birds should have crests trimmed to allow clear vision. Special attention is also needed to prevent lice or other vermin in crests, and regular dusting with insect-powder is advisable.

In winter weather colds can be prevented by using drinkers which avoid dipping the crest; and the breed is so heavily feathered that trimming of vent-fluff is recommended to ensure fertility.

Double-mating is not usually practised, though it is certain that Golds and Silvers would be rapidly improved in markings if two pens were used—possibly Chamois also.

At present Whites and Chamois are best and smallest. Size has increased rapidly in other colours recently—probably due to crossing with large breeds.

Suggested weights: Males 24–28 oz.; Females 18–24 oz. (Some colours greatly exceed these weights.)

The most serious defects are: Comb (if any) other than horn type; absence of muffling in all but white-crested colour varieties; split or twisted crest; legs other than blue or slate.

ROSECOMBS

Rosecombs (particularly Blacks) were formerly more highly-developed as show birds than any other breed. Probably no bird was ever exhibited, in large or small breeds, to exceed a good Black Rosecomb cockerel in exhibition style—though it must be admitted as truth that the excellence of his headpoints was frequently more due to surgical effort than to breeding!

They are not the easiest breed for a beginner—lobes go wrong in both winter and summer; they blister and wrinkle and acquire ruinous scabs. Rearing, too, isn't as easy as with larger, hardier breeds. Chicks require care in the early stages, and to bring them on for strong competition is the work of an expert. They lay very well in spring and summer, but are not suited to a damp, heavy soil, needing special housing when conditions are severe.

The breed exists in two colours, Blacks and Whites; but occasionally Blues have been bred. Blacks are by far the most popular, but it is regretted the charming Whites do not attract more breeders.

The ideal Rosecomb cock is a little gem, with comb broad and full in front, tapering in width to a long spike or leader at back, the leader set on firmly, perfectly straight and tapering to a point. Top of comb level and full of work (crowded with little fine spikes). Comb rises slightly from front to rear, the leader rising at the same angle (not, as usually stated, with an upward tendency from the comb). Comb must not be hollow on top or leafy in front.

The skull is broad and short, with short beak stout at base. Eyes full and bright, face fine in texture, earlobes absolutely round with rounded edges, of uniform thickness throughout and not hollow or dished, kid-like in texture and firmly set on the face. Wattles neat, round and fine.

Figure 20. Black Rosecomb Female. A Royal Show Champion for Mr. J. D. Kay

(Lobes are supposed to be standardised for cocks at a maximum size equal to a five pence piece, with those of females not larger than a half-pence piece; but these sizes would be too small for the modern show-pen. Nowadays lobes in cockerels sometimes exceed the size of a ten pence piece; which means that they may be exquisite for one or two shows, but rapidly spread to white in face, and become useless for show in a few weeks.)

Neck should be short, with plentiful hackles falling over shoulders and wingbows and reaching nearly to the tail. Shoulders and back broad and flat, chest broad, carried forward and upward with a bold curve. Stern should not taper off to nothing at junction with tail, and should have profuse long saddle-hackles.

Tail should be large, main feathers, sickles and furnishings long and broad, with rounded ends, not pointed. Sickles should curve in circular style almost to ground level, with side hangers broad, round-ended and plentiful. The neck, back and tail should form one long, continuous and graceful sweep in the form of a triple curve. Wings are large and strong but not unduly long, carried low to show front of thighs only.

Legs should be set well apart, with short, well-rounded shanks and four toes, straight and firmly spread. Carriage should be lively, graceful and jaunty, with the appearance of a bird ready to crow at any moment.

The female should be generally as described for the cock, allowing for usual sexual differences. Earlobes should be of good size but of course considerably smaller than in males. Tail should be of rather gay carriage. Years ago the tail was required to be well-spread, but nowadays it should be carried close though not tightly whipped.

Combs, faces and wattles are bright red in all varieties, free from white, which ofen appears on faces in large-lobed adult birds. Lobes should be perfectly white all over, preferably also under the rounded edges.

Figure 21. Pair of Black Rosecomb Bantams

Blacks should have sound rich black plumage throughout, with bright green sheen from head to ends of sickles, wing bars having extra bright sheen. Main tail feathers with bright green sheen are a point of rare excellence. Eyes should be brown or hazel, beak, legs and feet black, toe-nails white.

Whites should be of snow-white plumage, without straw tinge or false colouring, particularly on saddle and wings. Eyes red, beak, legs and feet white, without the bluish tinge that often comes from a mixture of Black blood. Whites in former days were bred with an exceptional flow of feather, greatly exceeding that of Blacks; but lobes were usually not so good, and head-points were inferior—frequently coarsened by the necessity for washing and drying.

DOUBLE-MATING?

Whites need only one pen to produce show specimens in both sexes, but Blacks have for many years almost invariably been double-mated. For producing first-class black cockerels use a male of show character with sound black legs, rich colour and brilliant sheen. Mate him to females with large, broad-fronted combs, very long leaders, and long tails showing a decided curve to the main feathers. It used to be said that hens for cock-breeding should be the deadest-black colour possible—sooty in fact, and without sheen; but this frequently means using females with purple tinge, which are undesirable.

For good pullets mate a small cobby cock, big in lobe, extra short in back, low on leg, and very rich in colour and sheen to females with exceptionally brilliant beetle-green colour throughout. These females will in fact be of exhibition character; and if the male with which they are mated has bronze or red in neck and back he will produce extra green sheen on his female progeny. If he is also somewhat hen-feathered in tail he will be a treasure.

Good birds can be bred, in both sexes, from one pen if care be exercised; but remember that if you want rich green sheen you have to use plenty of it in the breeding pen. An excess of green sheen may result in some males having red in hackles or back, but these will produce your richest-coloured pullets.

Don't forget that though dark legs are desired in Blacks, leg colour pales with age; and you won't get big thick lobes unless you breed for them on both sides.
from white, which often appears on faces in large-lobed adult

The Poultry Club gives the acceptable weights as: Cock 20–22 oz.; Hen 16–18 oz.

Serious defects include narrow build, hollow-fronted or leafy comb, tightly-carried wings, narrow feathers, blushed lobes, false colour in plumage, white in face; faulty colour in legs; and in Blacks, grizzled flights and purple sheen or barring.

Birds should be disqualified if shown with cut combs or mutilated faces, artificial colouring, and altered, removed or added feather.

SEBRIGHTS

This breed is a genuine bantam and is our oldest recorded British variety, having been evolved about 150 years ago. It exists in two colours, gold and silver—the latter by far the more popular. They are good layers of rather small eggs, but not easy to hatch and rear. Chicks require dry, well-shaded runs protected from cold winds.

The breed is classed as a henny—that is, males are hen-feathered, without curved sickles or pointed hackles. One description therefore suits males and females almost completely.

In both sexes the skull should be small, with beak short and slightly curved; comb rose-shaped, broad and square in front, firmly set on, free from hollows but full of work (covered on top with coral-like points) tapering off to a distinct leader or spike at rear. (It is frequently stated that comb and leader should be helmet-shaped, but the club standard particularly specifies that the spike or leader should be slightly upturned.) Eyes should be full, face smooth, earlobes flat and unfolded, wattles well-rounded.

The neck should taper, the cock's well arched and carried well back, the hen's upright.

Sebrights should have a very short back and compact body, with broad and prominent breast, low wings, and square tail, carried high and well-spread. (*Note*—the male is hen-feathered, without sickles, saddle hackle, or true pointed neck-hackle.)

Legs should be short, with slender shanks and four toes well spread. Carriage is strutting and tremulous, on tiptoe, and with an action resembling a Fantail pigeon. Plumage is short and tight, feathers not too wide but never pointed. Desired type of feather is described as almond-shaped, to avoid suspicions of pointed tips by breeding for almost square or blunt ends.

In Golds the beak is dark horn, in Silvers dark blue or horn. Eyes are black, or as dark as possible. Comb, face, wattles and earlobes dark purple or dull red, legs and feet slate blue. (*Note*—although the mulberry-coloured face is usual in females, in males it is nowadays impossible to attain—probably through breeding for exceptionally fine lacing over many years. Cocks should, however, have a damson eye-cere or surround—the darker the better.)

Plumage in Golds should be uniform golden bay with glossy green-black lacing and dark grey undercolour, each feather evenly and sharply laced all round with a narrow margin of black.

Figure 22. Mr. C. A. Parker's Gold Sebright Hen. Winner of
Many Prizes

In Silvers the ground colour is silver-white, otherwise
description of Golds applies.

Faults to avoid are shaftiness in Golds (pronounced pale
shafts to feathers), creamy-coloured wings in Silvers, black peppering inside tails, and coarsely-laced ends or spangle-tips to main
tail feathers.

In recent years uneven lacing (heavier on tips than on sides
of feathers), broken lacing at tips and thumb-nail lacing (not
right round the feather) have become usual—also a fine fringe of
ground colour outside the lacing on breasts. These should all be
stamped out; but if males come with a very little curve to main
tail feathers (slightly resembling hackles) they will usually be
good breeders.

Modern Sebrights have lost much of their old-time jaunty,
short-coupled, cobby type—probably through breeding consis-
tently for narrow, almond-shaped feathers. Other results of

55

modern breeding for fine lacing are loss of rich black and green (too frequently lacing being dull and brownish, though exceptionally fine) and lack of markings on flights. The breed is, however, of exceptional charm, and I know of no other which so attracts beginners.

Although many good specimens are bred in both sexes from one pen, it was usual formerly to mate two. Probably modern methods of single-mating are responsible for some of the faults previously detailed.

For cockerel-breeding use a soundly-laced cock inclined to heavy markings, with good breast and tail, well-laced in flights, and damson eye-ceres. Mate him to finely or lightly-marked females, but sound in black (not dark brown) and with good mulberry faces. Flights and tails should be laced to ends, the tails clear of peppering, smudging or spangle-tips.

For breeding good pullets the cockerel should be finely but completely laced, mated to heavily-marked females as dense in black as possible, but evenly and clearly laced down thighs, coverts and tails.

Small cocks and large hens are better than the reverse; and always inbreed as much as possible consistent with safety. If introducing outside blood do it carefully and in small doses; but when chicks become difficult to rear the only remedy will be an outcross.

Don't overdo fine lacing. Heavily-marked tail-tips are better than loss of wing-lacing, weak shoulders and lack of density in black.

Experience in breeding Sebrights is often claimed to have been the first introduction to sex-linkage methods, both in respect of Gold and Silver matings and the effects of black lacing.

Club standard weights: Cock 22oz.; Hen 18 oz.

Serious defects include—sickle feathers or pointed hackles on males; feathers on shanks; legs other than slate-blue; other than four toes; any deformity.

Single combs or similar lack of breed characteristics would of course merit birds being passed completely.

FAVEROLLES

The Faverolle is a beautiful bird. With his muffled face— like a "grandfather beard"—and feathered legs, coupled with a majestic carriage the male bird presents a handsome picture.

A full, broad breast is essential and the body should be heavily built, full and thick. Legs should be short and stout with five toes, the fifth toe being on the leg, turned up not touching the ground.

Figure 23. Mr. J. L. Milner's Salmon Faverolle Female.
A Consistent Winner

Colours are:

Ermine: White is the predominant colour with black stripes in hackle, black in wing flights and tail black.

Salmon: This is a mixture of colours—the male has a straw hackle; black muffle; back, shoulders and wing bows bright cherry mahogany; breast and thighs black. The female has breast, thighs, beard and muff of a creamy texture, the rest is a wheaten brown with darker shade stripes in the hackle.

White: Self white throughout.

The recommended weights are male 2½ to 3 lbs. and female 2 to 2½ lbs.

Serious defects are lack of full breast; faulty comb; wry or squirrel tail; white ear-lobes; lack of beard, muffling or foot feathers; other deformities. There should be five toes on each foot and any variation will call for disqualification.

Note: The description given here is necessarily brief and for a fuller explanation reference should be made to *British Poultry Standards.*

HOUDANS

Houdans have been bantamised around 30 years. They are little known, but are described because they are exhibited at important shows

Possibly they may not become popular, because, in spite of their crests, they have not much gaiety of appearance and are of utility character. They are, of course, miniatures of the well-known French breed, and somewhat resemble mottled Polands except that they have leaf combs and five toes on each foot.

The head of the Houdan is large, with a protuberance on top of skull, from which springs a compact, full crest, round on top and not split or divided, composed of plumage resembling hackle feathers, set so as to expose the comb and not obstruct the sight. The beak is short and strong, with wide nostrils; eyes bold, comb butterfly or leaf-type fairly small and balanced. Face bearded, the muffling large, compact and full, almost hiding the face and completely hiding the earlobes. The small round wattles are also almost hidden under the beard.

The neck is medium in length, with abundant long hackles; the body broad and long, resembling the Dorking in shape. Wings are carried close, tail full, with long well-curved sickles and furnishings.

Figure 24.   Houdan Male. A Winner for Mr. T. F. Bower

Legs are short and stout, set well apart and free from feathers on shanks. There must be five toes (another relic from a part-Dorking ancestry). Carriage is bold and lively.

With the exception of the crest (which is globular) the hen resembles the cock, allowing for normal sex differences.

Colour points: Beak horn, eyes red; comb, face and wattles bright red. Earlobes white, or white tinged with pink. Legs and feet white, mottled with black or lead-blue.

Plumage glossy green-black, with pure white mottling evenly distributed throughout, except on secondaries and flights. Sickles and tail coverts in cocks are irregularly edged with white. In young birds black predominates, but plumage becomes gayer with age.

There are no breeding problems and the standard has been drawn up to encourage single-mating.

Recommended weights: Male 24–28 oz.: Female 22–26 oz. Some tolerance in weight is advisable at present.

Serious defects include red or straw-coloured feathers; spurs outside the shanks, feathers on toes or shanks; other than five toes on each foot; loose crest obstructing the sight, or any deformity.

CHAPTER 5

MINIATURE LIGHT BREEDS

ANCONAS

ANCONAS ARE OF Mediterranean character, and miniatures of large Anconas. There is no bantam club to cater for them, but they are fostered by the large breed club.

Very similar in build and style to Black Leghorns, their plumage is tipped with white. They are attractive and extremely good for utility, but are very wild and flighty by nature. Approach their breeding pen incautiously, and they are "on the wing" at once. Wire netting ten feet high will hardly enclose them if they mean business, so clip three or four outer flight feathers (on one wing only of course) to throw them off balance. Alternatively, a more convenient approach, would be to place netting over the top of the pen.

Chief points in males are the usual Mediterranean single, upright comb, medium size with 5 to 7 broad, even serrations forming a regular curve on top, the comb following line of neck and free from thumb marks or excrescences; the head carried well back, eyes bright and prominent, face smooth and fine. Earlobes are medium size, almond-shaped, free of folds, and wattles are long and fine.

The long, profusely-hackled neck sits on a broad, compact body, with moderately long back tapering to the saddle, and full breast carried well up. Wings are large and well tucked, tail full, carried well spread, with flowing sickles and plentiful furnishings.

Carriage is active and alert, on legs of medium length, set well apart, the shanks strong and free from feathers, and the four toes long and spread. Thighs should be almost hidden by body-feather. The hen resembles the cock in general characteristics, but the comb falls gracefully to one side of the face, without obstructing the sight.

Figure 25. Mr. W. G. Groucott's Ancona Female. Winner of Many Prizes

Colour points are as follows: Beak yellow shaded with horn or black, eyes orange-red with hazel pupils. Earlobes white, face, comb and wattles bright red, the face free from white. Legs and feet yellow, mottled with black—the more evenly mottled the better. Wholly yellow beaks or legs are not desirable.

Plumage should be beetle-green with V-shaped white tippings throughout (the more evenly tipped the better) and with no splashing or lacing; the white tips clear of black or grey streaks. Feathers should be black to the roots.

NO BREEDING PROBLEMS

Except that headpoints can be improved by a modified system of double-mating there are no special breeding problems; and even this small necessity can be avoided by using, in the one breeding pen, some females with "cocky" heads and some with combs of exhibition character, as described under Andalusians.

A rosecombed variety is standardised in large Anconas, but is seldom seen in bantams.

Bantam weights, as recently revised by the large breed Ancona club, are—Male 20–24 oz.: Female 18–20 oz. (Weights recommended by the B.B.A. Standards Committee are slightly heavier).

Serious defects include white in face; light undercolour, plumage other than black and white; crooked toes, knock-knees; squirrel or wry tail. Roach back or other structural deformity to disqualify.

ANDALUSIANS

Andalusians, being from the Mediterranean group, have much of the typical Mediterranean shape common to Leghorns, Anconas, and Minorcas, leaning mainly towards Minorca character. The comb is single, upright, medium size, deeply serrated with spikes broad at base, the blade of comb following the neck but not touching it. It should be free from sidesprigs or thumb-marks. (In the female the comb droops as later described.)

The face is smooth, earlobes almond-shaped, flat, not dished or hollow, fitting close to the head; and wattles are long and fine.

The neck is plentifully hackled, the body long, with broad shoulders tapering to the tail. Breast is full and round, plumage compact and close, with wings well tucked up, ends hidden under saddle-hackle in the male. The tail should be large, with flowing sickles, not fan-shaped but carried moderately high.

Carriage is upright, bold and active, on longish legs, with shanks and feet free from feathers and four straight toes well spread.

The female resembles the male except for usual sex differences, but the comb should fall with a single fold to one side, partly covering the eye. (Note the slight difference in this respect from other Mediterranean breeds.)

Colour details: Beak dark slate or horn. Eyes dark red or red-brown; comb, face and wattles bright red. Earlobes white, legs and feet dark slate or nearly black.

Plumage clear slate-blue, each feather distinctly and clearly edged with black lacing of medium width, except sickles in males, which are dark or black; and the hackles in both sexes are glossy black, the hen's showing broad lacing at base.

Figure 26. Andalusian Male

There are no severe double-mating problems in Andalusians, and they would therefore be a good variety for beginners. Good specimens of both sexes can readily be produced from one pen, especially if two types of females are used.

Figure 27.   Andalusian Female
Both Andalusians are a little oversize, but indicate the correct type

The only item really meriting double-mating (or a modified form of it) is the comb. It is, however, readily possible to use, in one breeding pen, some females with exhibition-type combs and some with combs that are firm-based and nearly upright, somewhat masculine in character. This avoids separate pens, and gives quite good enough results, because Andalusians have medium combs without exaggerations.

It should be emphasised that so-called Andalusian blue in large breeds is not correctly blue, but a diffusion of black and white. Only about 50 per cent breed true to colour, the others being 25 per cent each of blacks (or near-blacks) and splashed whites. If the blues produced are mated together, blue to blue, approximately the same proportions of blues, blacks and splashed-whites will be produced; but it has proved almost impossible to convince breeders (either beginners or old hands) that by mating the black offspring to the splashed whites all their progeny will come blue.

When these blues are once again bred together, blue to blue, the same cycle occurs, with the usual proportion of blacks and splashed-whites.

Recommended weights—Male 24–28 oz.: Female 20–24 oz. These weights are recommended by the Poultry Club.

Serious defects include white in face, red in lobes, white in feather, sooty ground colour; lopped comb in males, upright comb in females; red or yellow in hackles, or any deformity.

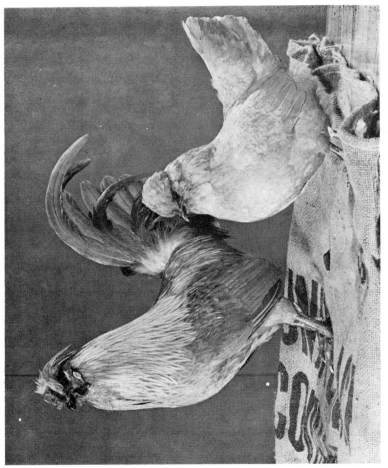

Figure 28. Mrs. W. Roxburgh's Prizewinning Pair of Araucanas
(Photographer: Michael J. Bone)

ARAUCANAS

The Araucanas are beautiful birds and deserve more attention from breeders. Although classified as a light breed in bantams it might be better to regard them as "ornamental". A notable feature is the fact that the egg is blue or green—this often receives too much attention, thus giving the breed a "freak" tag which it does not deserve.

The body should be long and deep, yet free from heaviness. On the small head there should be a compact crest and thick muffling and ear muffs. The comb should be of the pea type of an irregular shape.

Legs should be of medium length, with shanks free of feathers, and with four toes. They should be willow, olive or slate, and eyes should be dark orange.

Colours are Black Red, Birchen, Brown Red, Gold Duckwing, Cuckoo, Pile, Black, White, and Lavender. These tend to follow the conventional pattern most being similar to the colours for Modern Game.

Recommended weights are Male 26–30 oz. and Female 24–28 oz.

Serious defects are as follows:
1. Cut away breast or roach back.
2. Tail not at 45° angle (wry or squirrel).
3. No muffling or crest (the latter should be *compact* but not large).
4. Faulty comb (should be minimal in female).
5. Wrong colour eye (e.g. pearl) or legs.
6. Low wing carriage.
7. Faulty colouring.
8. Other deformities or defects such as feathered legs.

LEGHORNS

These are miniatures of the old-established Mediterranean breed which years ago set the fashion in winning laying contests. The Leghorn Bantam Club caters for the breed, which is now gaining popularity. As would be expected the breed should rank as first class layers, thus combining beauty and utility.

The breed presents capital openings for beginners, with choice of numerous colours and markings, of which Whites are best for size and headpoints. Browns are probably better for type, but too often fail badly in colour and lobes.

Leghorn bantams are of the Mediterranean group, and therefore should have large single combs (erect in the male), and white or cream-tinted earlobes. The comb should be evenly serrated, with spikes broad at base, and should extend well over back of head with blade following, but not touching, the line of neck.

Earlobes should be kid-like in texture and pendent (not round), wattles long and thin. Comb in the hen should droop over the face without obstructing the sight and when opened out (allowing for flexibility) should resemble a male comb in shape. Females with non-drooping combs should not be exhibited. The place for a hen with upright comb is the cock-breeding pen, not the show cage.

Figure 29. Mr. A. Howard's White Leghorn Male. Winner of Numerous Prizes

The head points described, with flowing feather and typical wedge-shaped Mediterranean build, are the chief breed characteristics. The skull should be fine, and comb should not project in front as far as the beak. Rosecombs are also standardised in the large breed, but not yet seen in bantams. Single combs should be free from thumb-prints and side-sprigs.

The neck is long and profusely-hackled, body wedge-shaped in profile, due to strong abdominal development. Shoulders wide, narrowing towards rear. Breast round and prominent, back long and sloping down towards tail, which is moderately full, flowing, and carried at an angle about 45 degrees from the back. Wings large, carried close and tucked up.

Legs are moderately long, shanks fine and round, free from feather. Flat shins and split shanks are objectionable. Toes number four.

Plumage is silky and free from excess in fluff. Carriage sprightly and alert, with no trace of stiltiness.

Except for the tail (which is carried closely and at a lower angle) and the drooping comb, the female resembles the male, allowing for sex distinctions.

Colour points are: Beak yellow, or yellow and horn in the darker colours. Eyes red, comb, face and wattles bright red. Earlobes pure opaque white or cream—white preferred. Legs and feet yellow or orange.

In Blacks the plumage should be rich black with plentiful green sheen free from any other colour, with undercolour as dark as possible. As in all black breeds with yellow legs, this means double-mating (see separate chapter). Exhibition males and females cannot be bred from one pen.

Figure 30.   White Leghorn Female. A Top Class Bird from Mr. A. Howard

Whites can be bred by single-mating methods, though the best head-points are produced by using separate pens for breeding show cockerels and pullets. In the cock-breeding pen females with firm-based (and possibly upright) combs are used, and for breeding pullets a cock with floppy, weak comb is mated to exhibition-type females. Plumage should be pure white, free from straw tinge or other false colour.

In Browns the neck-hackle of the cock is rich orange-red, shading off at tips to pale lemon colour, but crimson-red at front

67

below wattles; each feather striped with black. Wing bows, back and shoulders deep crimson-red or maroon, wing coverts steel-blue. Primaries brown, secondaries deep bay on outer web, black on inner. Saddle rich orange-red, shading off to match neck-hackle as nearly as possible. Tail black with rich green sheen; any white in tail is very objectionable. Tail coverts are black edged with brown; breast and underparts glossy black, completely free from brown markings or splashes.

The hen's neck hackle is rich golden-yellow, broadly striped with black. Body colour is soft rich brown, very closely and evenly pencilled with black, free from shaftiness, and wings free from red or rust. Breast is salmon-red, approaching maroon near head and wattles, and shading off to ash-grey at thighs. Tail black, outer feathers pencilled with brown.

Apart from the frequent use of double-mating methods for headpoints, colour-descriptions in male and female standards make it impossible to breed really good show specimens in both sexes from one pen. For description of breeding for colour and marking see special chapter.

Blues are occasionally seen and should be of even light-blue tint, unlaced. No straw tinge or brassiness permissible, but hackles are of darker colour in males.

Other colours are standardised in large breeds but seldom seen in bantams, though occasionally Duckwings and Cuckoos are shown. Colours in such varieties follow game standards fairly closely.

As noted under the various colours, the breed is very definitely one which calls for double-mating. Except perhaps in whites and blues, very little success would attend attempts to produce both show males and show females from one pen. In any case, headpoints in all sub-varieties are greatly improved when double-mating is practised.

To produce good combs on cockerels, mate a male with perfectly erect, firm-based comb of exhibition character to females with combs that are upright, or nearly so, firm-based and "cocky" in style. For pullets, mate hens with exhibition head-points to a male that is thin and weak in comb; preferably one whose comb droops like that of a good female.

(*Note*—It is not possible to produce on females the standard type of comb, falling gracefully to one side of face, without obstructing the sight, and at the same time retain large size, unless the comb first folds slightly to one side, then falls the other. A single or simple droop to comb means unavoidable obstruction of sight in one eye unless the comb is of American type and much smaller than British practice.)

Weights recommended are—Male 32 oz.; Female 28 oz. It should be remembered that these are miniatures of a small, light breed and therefore some may think that these weights are excessive. There is no justification for weights outside the comparable breed-group.

Serious defects include lopped combs in cocks, or erect combs in hens; combs obstructing sight, earlobes red, white in face, legs other than yellow or orange, wry or squirrel tail, any deformity. Also in Blacks, dark legs or eyes; and in Browns, white in feather.

MINORCAS

For many years Minorcas have been amongst the best of all bantams for laying; and in some strains they have been developed of excellent breed character and moderate size. Minorca bantams of moderate size have always proved capable of laying large eggs and plenty of them—probably more than any other breed of like size.

The breed follows general type in Mediterranean fowl; and being a miniature of the largest Mediterranean breed may legitimately be somewhat heavier than Leghorns, Anconas and Andalusians.

Chief characteristics are the typical Mediterranean head-points, which include a large, single comb (vertical in the cock, drooping in the hen) moderately rough in texture, with broad based spikes forming even serrations (five preferred); blade of comb following line of neck at rear without touching it, and front of comb not projecting beyond beak. Comb free from thumb marks, twists or side sprigs. Eyes are dark—the more nearly black the better (males frequently fail in this respect).

Figure 31.   Mr. Ken Binn's Award-Winning Black Minorca Female

69

Earlobes are required to be smooth and flat, elongated, not round, widest at top, kid-like in texture, close-fitting, not dished. (*Note*—size of lobe is not fixed, but in the show-pen large lobes are the rule, though not demanded by standard.) Face should be unwrinkled and hairless, showing no white; wattles long and rounded at ends.

The graceful neck is long with plentiful hackle, the body broad at shoulder, compact but fairly long in back, with deep straight keel. Breast full, body wedge-shaped in profile, with strongly developed abdomen in females. Wings moderately long and close fitting. Tail full. with long, broad, curved sickles carried well back.

Legs and thighs are medium in length, shanks strong, free from feathers and well apart, without knock-knees or stiltiness. Toes number four. Carriage is horizontal but graceful.

Males and females are similar in character, allowing for sex-differences; but the hen's comb should fall gracefully over one side of the face without obstructing the sight. Upright or nearly-upright combs in females should be discouraged.

Figure 32.   Mr. Raymond Shaw's Prizewinning Black Minorca Female

The Black has dark horn-coloured beak, dark eyes (nearly black); face, comb and wattles blood-red. Earlobes pure white, legs and feet black (or very dark slate in adults). Plumage black throughout, with brilliant green sheen.

In Whites the beak should be white, legs and feet pink-white; eyes red, comb, face and wattles blood-red, ear-lobes pure white, plumage lustrous white.

Blues have been standardised in the large breed, but are not yet seen in bantams.

Except for headpoints there would be no need for double-mating in Minorcas. Exhibition birds of both sexes could well be produced from one pen. It is, however, because too many breeders attempt this that females are so often shown with combs stiff and upright, or with too little fall or droop.

Standard headpoints can be produced much better by mating two pens—the cockerel-breeding pen having a good-headed male, mated to females with firm-based, upright (or nearly upright) combs "cocky" in character. For pullets a male with drooping, floppy, thin-based comb is mated to females with exhibition headpoints.

Except for these remarks there are no special mating problems. Simply select for breed character, using birds with long backs, dark eyes, and good shanks, and avoiding grizzled flights and crooked breasts.

Minorca bantams are not only charming, but useful. Probably no other breed suits a back-yard fancier so well.

Maximum weights prescribed by the large breed club (which also caters for bantams) are—Cocks 34 oz. : Hens 30 oz.

Smaller specimens should be favoured, other points being equal. This is a sound specification. Recommended weights could well be slightly smaller.

Serious defects include white in face, blushed lobe, wry or squirrel tail, feathers on shanks, foul colour in plumage, incorrect leg colour, other than four toes, crooked breast bone, any deformity.

HAMBURGHS

There is great beauty in the markings of Hamburghs, which exist in two colours in each of the two varieties—Golds and Silvers in Spangled and so-called Pencilled markings. The breed is somewhat on the lines of a Rosecomb, but larger and not so refined. The existence of the Black Rosecomb is, in fact, the obvious reason why Blacks do not exist in Hamburgh bantams, though standardised and popular in the large breed.

Size can always be reduced by inbreeding, using birds that obviously possess the necessary genetic factor; but colour, markings, and wealth of feather are main considerations.

The skull of the exhibition male is fine, with short, well-curved beak, bold and full eyes; rose comb of medium size, square in front, firmly set, tapering to a long, fine spike or leader, straight and without tendency to droop. Top of comb free from hollows, and full of fine work or small spikes. Face smooth, lobes round, smooth, flat and undished. Wattles round and fine.

The neck is of moderate length, with plentiful hackles covering the shoulders. A moderately long but compact body shows flat back and wide shoulders tapering towards tail, with well rounded breast and closely-tucked wings; while the long sweeping tail has broad sickles and plentiful furnishings, all carried well up.

Legs are medium in length, with slender thighs and fine round shanks free from feathers, and with four slender toes well spread. Carriage is alert, graceful and bold.

General characteristics of the female are similar to the male, allowing for normal sex distinctions.

In all Hamburgh sub-varieties the beak should be dark horn, legs lead-blue, earlobes white; eyes, comb, face and wattles red.

In the Gold Pencilled variety the cock is bright red-bay or golden chestnut, except the tail, which is black, with a narrow gold edging or lacing right round sickle feathers and coverts.

The hen has ground-colour similar to the cock. Neck hackle should be as clear of markings as possible, all remaining feathers distinctly and evenly "pencilled" straight across with fine parallel lines of rich glossy green-black; barring or pencilling and ground-colour each the same width, and the more markings the better.

Figure 33. Mr. J. D. Kay's Champion Silver Spangled Hamburgh Pullet

The Silver-Pencilled variety is similar to Golds, except that ground-colour and lacing are silvery-white.

In Gold-Spangled males the ground-colour is rich mahogany or bay. Markings, spangling, tipping and tail are all rich green-black. Hackles and back are striped with black down the middle of each feather. Breast and underparts are tipped throughout with round green-black spangles or spots, small at throat and

72

increasing in size towards the thighs; spangles as large as possible without overlapping. (Frequently, exhibitors "thin" feathers to comply with this requirement, but the practice should be prohibited.)

WING MARKINGS

Wing bars are formed by two distinct rows of large round spangles, parallel across each wing, with gentle curves. Secondaries are similarly tipped, and their diminishing lengths form "steppings" on upper edge of wing. Wingbows show dagger-shaped tips of rich green-black.

The Gold-Spangled hen is similar generally to the cock. Tail coverts are black, with sharp edging or lacing of gold. Remaining feathers tipped throughout with round green-black spangles, commencing high up the throat, and as large as possible without overlapping.

The Silver-Spangled variety is similar to the Gold except that ground-colour is silvery-white, and that in both sexes the tail is white, tipped with bold green-black "moons" or spangles, the coverts being similarly marked.

A peculiarity worth remark is that whereas in Silver-Spangles the tail is silver, in Gold-Spangles tail colour is black, instead of resembling the general ground-colour.

Figure 34.   Silver Spangled Hamburgh Cock. Winner of Many Awards
for Mr. J. D. Kay

73

It may be considered a disadvantage that the breed requires double-mating. It is not possible regularly to produce really good exhibition birds of both sexes from one pen. In general many birds shown recently have been produced by single-mating, principally because many fanciers are not cognisant of old methods.

The very moderate quality of most present-day exhibits reflects the need for more skill in breeding. Some excellent examples of pullet-marked males have been exhibited recently at shows. Perhaps the acceptance of this stamp of bird as the standard for males might avoid the necessity for mating two pens. It is worth considering. Many people (myself amongst them) hold the view that the markings of a pullet-breeding cockerel are more beautiful, and more in keeping with breed character, than the standard male.

Silver-Spangles are considerably the most popular. For cocks mate a good show-type male of brilliant markings to females bred for many generations from a cock-breeding strain. These should have strong "cocky" headpoints, with neck hackle heavily ticked but white underneath. They will be well spangled elsewhere, particularly on shoulders, with strong wing-bars; tail long, clear, and solidly spangled. If main tail feathers show a curve, so much the better.

For pullet-breeding select females of exhibition quality and markings, and mate them to a "hen-feathered" male, which should be descended from several generations of pullet-breeders. He should carry no sickles, but should be very well spangled, and marked as much as possible like a female. In long-established pullet-breeding strains some males come with necks spangled like females. These are the kind to use.

If you wish to breed from one pen only, mate a really good show-type cockerel to two exhibition-coloured hens and two cock-bred females as formerly described; but if you want single-mating to be really successful you must carefully mark all progeny. Unfortunately this means trap-nesting.

Gold-spangled matings follow similar lines. In this colour males are not equal to Silvers but females are sometimes good.

In pencilled varieties the reverse applies, Golds generally being better than Silvers. Colour is usually fairly good, but markings are poor—possibly because Hamburgh men appear never to have been able to decide which is ground-colour and which is pencilling. (The so-called pencilling in this breed is a form of barring.)

Length of feather is of importance, a long flowing tail in males with plentiful side-hangers, being a necessity. If you want good cockerels, breed always from the best-feathered males and longest-tailed hens, preferably those with some curve to main tail feathers; and if they have flowing hackles so much the better. Faults in pencilling are not of importance in these females—in fact they can be of advantage.

74

The usual general methods of mating apply for pullet-breeding, remembering that for this you need exhibition females with first-class pencilling, mated to a male showing as much female character and markings as possible. These will only be produced from a well-established pullet-strain.

It is not possible, in a short article, fully to describe mating methods, but the principle involved is very simple—in multi-marked breeds mate, for pullet-breeding, exhibition-type hens to males with strong female characteristics, and vice versa. Experience will teach what you fail to learn from your club.

Recommended maximum weights—Male 24–28 oz.: Female 22–26 oz. (Pencilled Hamburghs often greatly exceed these weights.) Serious defects are red earlobes, white in face, squirrel tail, wry tail, or any other deformity.

SCOTS GREYS

There seems little reason why Scots Greys should have become almost extinct. Once they were very popular, and could be depended on to win in good company in Variety classes; but very few now exist, but they are being revived.

Plumage is clear steel-grey ground colour, with distinct narrow black barring across the feather. Legs should be white, preferably spotted with black; not yellow, dark or smutty. Comb, wattles, face and lobes are red—though lobes frequently show some white.

Figure 35. Prizewinning Scots Grey Cockerel. Bred by Messrs. Jas. R. and M. Smith

Figure 36. Messrs. Jas. R. and M. Smith's Scots Grey Hen.
Winner of Top Awards

General build and type are somewhat racy, with longer tail and sickles than in Plymouth Rocks. It is splended that they are being revived.

The photographs are very typical of these birds. Suggested weights are—Males 22–24 oz. : Female 18–20 oz.

Colour and markings are of prime importance, and the most serious defect is evidence of alien cross, or characteristics of other breeds.

WELSUMMERS

A specialist club exists to cater for the large and bantams of this breed which is Dutch in origin. It is considered, however, that Welsummer bantams should be judged for utility character as well as for breed points. These are docile, attractive birds well suited for those with limited space.

In the large variety the Welsummer was standardised by The Poultry Club as a light breed—the only one that lays brown eggs. Probably it will at some future date be transferred to a new medium-weight grouping; until then it might have been better to include them, in our bantam standards, amongst miniature editions of heavy breeds—because their characteristics resemble much more the sitting type. than the non-sitting light varieties.

The male has a wedge-shaped body with good depth and width, a broad breast and a fair length of back, with well-developed wings and tail carried high, as befits a bird of utility stamp. Carriage is active and alert, with legs neither short nor stilty, thighs and shanks strong and well apart, the four toes well spread. The neck is fairly long and the comb single, straight medium in size and erect, with five or six serrations. Earlobes are small, wattles long and thin, neck fairly long and well-hackled.

Figure 37.   Welsummer Pullet. A Prizewinner for Mr. P. Parris

Allowing for sex, the hen resembles the cock, but should be particularly plump in breast and well-developed in abdomen, indicating laying-power.

The Welsummer has a yellow beak, eyes red or deep yellow; comb, face and earlobes bright red, legs and feet deep yellow to orange.

Male plumage generally is a rich golden brown; breast and thigh-fluff black, mottled with red. Neck hackle is deep red-brown, free from striping. The tail is black with beetle-green

77

sheen, peppered with brown on outer feathers; wings red-marked on shoulders, flights black. The general colour-scheme is that of a modified utility-type black-red.

The female is a soft rich brown, peppered with black partridge markings, with distinct light shafts to feathers. The breast is salmon, rump and thighs red-brown. Neck hackle is red-brown with golden-orange fringes, free from black striping. Tail black, edged at base with brown. Wing primaries have black inner webs, brown outer. Secondaries have brown outer webs peppered with black, inner webs black peppered with brown.

There are no special breeding problems, the standard being drafted to avoid any points likely to require double-mating. Good specimens of both sexes can be bred from one pen, thanks to the standardising of mottled breasts in males and the avoidance of striped hackle and lacing.

Concentration on type is therefore the only necessity, with upstanding, alert character. Unless, however, the deep brown egg factor is improved they may not become really popular.

Serious defects include white in earlobes, self-coloured breast in males; side-sprigs, comb other than single, willow legs, feathered shanks or any deformity.

CHAPTER 6

MINIATURE HEAVY BREEDS

AUSTRALORPS

AUSTRALORPS ARE REALLY miniatures of old-type Black Orpingtons. In show bantams there is often practically no difference between them, except that frequently Australorps are considerably too high on leg, though Orpingtons do not usually approach the large breed in low build. Doubtless the legginess of Australorp bantams is the result of an endeavour to keep the two breeds distinct. They are otherwise excellent in breed character, feather and colour.

The cock's general characteristics greatly resemble the Orpington, as might be expected, the comb being small, erect and evenly serrated, eyes bold and full, standing well out and high in the skull. The face must be smooth and unwrinkled, not fleshy, and without heavy eyebrows; earlobes small, wattles close-fitting and fine.

The neck should be fairly long, the body wide and deep, as in the Orpington, but reasonably long, with tight saddle rising symmetrically to tail; shoulders broad, breast full and deep, wings tucked up and broad tail not carried so high as in the exhibition Orpington.

Legs are *medium* length and well apart, with rounded pliable shanks, medium in bone, free from feathers, and with four well-spread toes. Carriage is alert, active and cobby. Short legs are not desirable, and short backs are not standard.

Plumage is short and close, avoiding round cushions, fluffy saddles, long tail feathers, and baggy thighs. Abdominal fluff should be silky and fine. Excess fluff and long sickles are undesirable, and the plumage should show a "fly-back" tendency when brushed the wrong way.

Figure 38. Mr. Raymond Shaw's Australorp Male. Winner of Many Best-in-Show Awards

General characteristics in females are as described for cocks, allowing for sexual differences.

Beak of the Australorp is black, eye black with black or very dark iris. Comb, face, wattles and earlobes red. Legs and feet black with whitish pads. Plumage rich black throughout, with very intense beetle-green surface sheen, free from purple or blue.

Figure 39. Australorp Female. A Prizewinner for Mr. Alan Maskrey

There are no special points in breeding, and no necessity exists for double-mating. In fact, the large-breed standard was specially framed to avoid it. The main points to consider, in mating stock, are that it is easier to breed shanks too long than too short; that white in wings, general plumage and under-colour must be avoided, especially if accompanied by white quills in flights; that purple or bronze sheen and barring are objectionable and hereditary; and that white in lobes is a grave defect that will repeat itself indefinitely once it is bred in. These defects are seen far more often in bantams than in the large breed —so are yellow pads and willow legs, which are evidences of an alien cross. Females often have gypsy faces which should be avoided.

Recommended weights are—Male 36 oz.: Female 28 oz.

Serious defects include white in wings, plumage or earlobes; bronze and purple sheen or markings; coarseness and excessive feather.

Very serious defects (for which birds should be passed) are willow legs, yellow pads, feathers on shanks or toes, side-sprigs or double-blade to comb, light or red eyes, and any deformity.

BARNEVELDERS

Barnevelder bantams have now so declined in popularity, possibly because of the loss of the brown egg factor for which the big breed is noted. Genetically, no known breed of bantam carries the deep brown egg factor, hence it is almost impossible to perpetuate it when a large breed is bantamised.

In bantams only one colour exists—the laced variety. The cock is of somewhat upstanding character, with fairly long neck profusely feathered, and single comb of medium size, well serrated. The body is compact in concave back, broad shoulders, deep breast, and full stern. Wings are short and carried high, tail full but not long.

Legs are medium length, fairly stout, unfeathered on shanks, with four well-spread toes. Carriage is alert and active, plumage fairly close and tight. The hen resembles the cock in general characteristics.

Beak is yellow with dark markings, legs and feet yellow, eyes orange, comb, face, wattles and earlobes red.

COLOUR CHARACTERISTICS (DOUBLE LACED)

In cocks the neck hackle is black with red markings, not regularly striped. The breast is black-laced on red-brown ground, the back deep red with wide black lacing.

Wings are red-brown with broad black lacing, bars red-brown; primaries and secondaries black with red markings. Saddle is also black with red markings. Thighs, tail and underparts black. All black feather and markings should have rich beetle-green sheen.

In hens the neck hackles are black, with or without brown markings. The breast is red-brown with broad black lacing; back and cushion red-brown, *double-laced* with intense black, the outer lacing being very broad. Tail is black, coverts red-brown laced with black; wings black with red-brown bars, primaries and secondaries black with red-brown outer edges. All black to have rich green sheen.

A "Partridge" variety existed but is not now often seen. There were not wide differences in colour, but the laced variety was more cleanly marked.

The Standard was deliberately so framed that good birds of both sexes could be produced from one mating. Thus no craze for double-mating to perfect lacing and marking is necessary.

Recommended maximum weights—Male 32 oz.: Female 26 oz.

Serious defects include white in lobes, side sprigs on comb; legs and feet other than yellow, feathered legs or feet; any white in plumage, any deformity.

ORPINGTONS

The best Orpington bantams are really typical, and **no other** breeds have quite similar character and style. The small **head** and comb, very deep body in comparison with length, distinctive saddle shape and carriage, and short shanks with thighs almost hidden (but positioned to give correct body balance) present a most attractive bird.

The wide rising saddle gives the back an extremely short, concave outline. The small wings are well tucked-up, ends hidden by an extraordinary wealth of saddle hackle; and the short compact tail is carried high.

Figure 40.   Mr. Will Burdett's Prizewinning Black Orpington Females

Carriage is graceful and active, with a general impression, to the eye, of greater depth than length; and some typical Asiatic characters should be present, such as a balanced, forward-tilted body. Plumage is extremely plentiful, but not quite as soft as in the Cochin though very well-fluffed. (The original standard for the large breed called for plumage fairly close in character; but this was soon over-ridden, and for many years they have been noted for a wealth and softness of plumage seldom seen in other breeds.)

An excellent description* deals with the breed under the following headings:

1. Shape of body—this should be *globular*.
2. Head and neck—the head should be small on a curved full neck not of excessive length.
3. Tail and legs—the broad tail should be a natural extension of the saddle cushion which starts in the middle of the back. Legs should be short and set well apart to give perfect body balance.
4. Carriage—should be graceful and active with "bold, noble dignity".

The aim is to achieve the perfect *type* without excessive size. Viewed from all sides the globular shape should be apparent and any birds with narrow chests or shallow keels should be penalised.

In Blacks the beak, legs and feet should be black; toe-nails, soles of feet and skin white. Eye black with dark brown iris; comb, face, wattles and earlobes bright red. Undercolour as sound black as possible, and plumage sound rich black with brilliant green sheen.

The Buff has white beak, legs, feet and skin, with red or brown eyes, red preferred. Comb, face, wattles and earlobes bright red. Plumage sound, clear even buff to skin all over, not showing two or three shapes, though male hackles are usually more lustrous than body colour. Quills buff throughout. White or black in tail and wings are objectionable.

The White should have red eyes and red headpoints; beak, legs, feet and skin white and plumage snow-white throughout.

Blues should resemble Blacks in all respects other than plumage colour which should be slate blue, dark across the top receding to medium, with dark lacing on feathers.

I have never seen any advantage in double-mating this breed, though no doubt in Buffs the old-time method of using a lemon-buff cock with hens of somewhat darker shade would produce the best cockerels; while pale-coloured show hens mated to a slightly darker cock would probably result in excellent

---

* *Poultry Club Year Book,* 1968, Orpington Bantams—Type—The Essential, Will Burdett.

pullets. Too great a difference in colour between the sexes would, however, lead to patchiness, abrupt differences in colour on wing, laced plumage, and mealiness.

Breed from birds with sound buff plumage to the skin, but don't mistake very pale buff for white. Undercolour can be pale and yet remain buff; but quills should be a distinct buff to the skin, and white quills in wing-feather should be avoided.

Buff hens usually get paler as they age, but Buff cockerels often get darker, particularly in tail and wing-butts.

When breeding Blacks take care to maintain rich green sheen, and to avoid bronzed or purple barring. You can't get enough green sheen unless you have it on both sides in the breeding pen; and if breeding for it results in the production of a cockerel with red feathers in hackle and wing don't despise him —he will breed pullets with really rich sheen.

Whites present no difficulties other than the usual necessity of special selection for short bodies, type, feather and carriage.

Recommended weights—Male 32–36 oz. : Female 28–32 oz. Blacks and Whites are already sufficiently reduced in weights to render much latitude unnecessary.

Serious defects include side sprigs, white in lobes; long legs, feather or fluff on shanks and feet; yellow skin or yellow in shanks and feet, coloured shanks in Buffs or Whites, coarseness in head; squirrel tail, flat back, square breast, split wing.

PLYMOUTH ROCKS

Three colours have been standardised in bantams (Barred, Buff and Partridge), though Blacks and Whites have sometimes been shown, and an occasional Columbian. These Blacks, Whites and Columbians have almost always been single-combed sports from Wyandottes.

The Barred variety in bantams probably evolved from the Scots Grey.

Characteristics of cocks in all colours include an erect single comb, moderate in size, with well-defined serrations, medium-sized wattles, and fine-textured lobes. Hackle is profuse on a well-curved neck, the back is broad and of medium length, and saddle feathers abundant but not unduly long. Breast full, deep and rounded; wings well tucked up, ends covered by saddle-hackle. Small rising tail, with curved sickles of medium length, and abundant coverts hiding the main tail.

Carriage is upright and alert, with thighs well apart and shanks of medium length, straight and stout, free from feathers. A tendency to knock knees must be avoided. Toes number four, straight and well spread.

The hen generally is similar to the cock, allowing for sex-variations.

Eyes are rich red or bay; beak, legs and feet bright yellow. Comb, face, wattles and earlobes bright red.

Figure 41. Mr. J. Shortland's Prizewinning Buff Rocks. A Majestic Pair

85

The Barred variety plumage has a white ground with bluish tinge, each feather barred across with black bands having a beetle-green sheen; the bands moderately narrow and equal in breadth, colours sharply defined. Barring should continue through the shaft and under-fluff, each feather finishing with a black tip. The whole appearance should be bluish, not brownish or brassy, and be of uniform shade all over. Barring is narrower in males than in females.

The Buff has plumage clear, sound pale golden buff throughout, with tail, wings and fluff in harmony, and without lacing, peppering or mealiness. If too pale the colour does not last for very long.

Barred Rock bantams are still somewhat more popular than Buffs, though the latter are probably better in type. Worst failings in the Barred variety are brassiness, dull brownish-grey appearance instead of blue-black, dark wing ends, and variations in type.

Good Barred birds can be bred in both sexes from one pen, but in the popular days of the large breed double-mating was almost invariably practised. In these more enlightened days, when the study of genetics has become part of a modern breeder's life, the problems that confused old hands would hardly worry us. They would be taken in our stride.

Old-time breeders evolved practical matings which weren't scientific, but which gave improved results. Briefly, for cock-breeding, rich-coloured hens of medium-to-dark shade were used. This method threw some black pullets, but gave a reasonable proportion of show cockerels.

For breeding show pullets a cockerel of light shade was mated to standard-coloured hens. This gave good females but washy, pale cockerels. The proportion of really well-marked specimens was never high, and it is unlikely that any known method will produce big proportions of winners while our standard requirements remain as now.

In single-mating, it was found that the best cockerels usually came from the darkest hens, and the best pullets from hens of standard colour. It was a long time, however, before breeders realised that mating a finely-barred male to their black female "sports" (mutations from Barred matings) would give good male progeny.

Breeding Barred Rocks is always somewhat chancy. Only personal experience with one's own particular strain will solve the problems involved. In any case, the bantams appear genetically to be entirely different from the big breed, and therefore, present different problems. It is necessary to retain the barring without losing the blue-black.

Buffs are a single-mating variety, needing only care in selecting for type and level colour. They are excellent layers of large eggs.

The Partridge variety is typical of this colour in other soft-feather breeds. The male has neck and body of black with the hackle edged with red as well as the back, saddle and wing bows being of brilliant red. The female is of a reddish bay colour, and the body feathers have a distinct pencilling of black.

Recommended weights—Male 26–30 oz.: Female 22–26 oz. Standard weights of the Buff Rock (large breed) Club and the Barred Rock Bantam Club are respectively slightly below and slightly above these figures.

Serious defects include indications of feather or fluff on shanks or feet, shanks other than yellow; white earlobes, pale or faulty-coloured eyes; slipped wing; foul-coloured feather in Barred variety; spotted, laced or mealy plumage in Buffs; black or white in wings, or white in tail in Buffs; any deformity.

RHODE ISLAND REDS

It seems unnecessary to stress the popularity of R.I.R. bantams. At shows they even exceed the large breed in numbers; and while they undoubtedly have a long way to go before they achieve comparable type and shape, in colour they can hold their own. There is little doubt that the general usefulness of the Red bantam (as in the large breed) is mainly responsible for its popularity.

This is one of the few bantams which carry a considerable amount of large breed blood. A number of prominent breeders have reduced them with varying amounts of bantam blood. It is difficult to accept, however, without a great deal of reserve, statements sometimes made that they have been brought down from the large breed without the use of bantams. There must have been at least one original bantam introduction, or how could a start have been made?

Figure 42.   Mr. A. Anderson's Champion Rhode Island Red Pullet

In general character the body should be brick-shaped or oblong, the back horizontal, and the short tail continued almost in line with the back, giving an even greater appearance of straight, level length. Wings are large, horizontal and well folded, and tail must not be of Leghorn character or have long flowing sickles; this is important. Birds with uncharacteristic tails should be passed in the show pen, or heavily penalised, otherwise the fault will be perpetuated.

The comb is single and upright, medium sized and straight, with five even serrations. Lobes are fine and well-developed, wattles medium in size. The neck is moderate in length, carried forward and profusely hackled.

The body is broad and long, its distinctly oblong shape accentuated by a long straight keel-bone extending well fore and aft. The broad and full breast is carried nearly perpendicularly in line with base of beak. Feathering should be close, moderately full in fluff but nowhere loose.

Carriage is alert, active and balanced, thighs large and well-feathered, shanks medium length and free from feathers, the four toes strong and well spread.

The female resembles the male in all essential particulars.

Colour: Beak, legs and feet red-horn or yellow; eyes red, comb, face, earlobes and wattles brilliant red.

PLUMAGE OF THE MALE

The cock's plumage should have hackles matching body-colour and without black markings. Wing primaries have lower webs black, upper red; secondaries red in lower web, upper black; flight coverts black, wing-bows and coverts red. Tail and sickles black or green-black; coverts mainly black, but red nearing the saddle.

The whole of the remaining plumage rich brilliant red, free from shaftiness, peppering, mealiness or "ginger"; under-colour and quill-colour red or salmon, without smut or white. Black or white in undercolour is most undesirable; and (other things being equal) the richest red undercolour shall carry the award.

The bird should have a very brilliant top-lustre giving a glossed appearance. The red now favoured is an extremely deep chocolate-red—a rich red toned down by infusion of black; and though some breeders disagree with this description, birds of lighter and brighter colour seldom receive prizes.

The hen should have hackles matching body colour, tips of lower feathers having black ticking but not heavy markings. Tail black or green-black, remainder as described for the cock, except that the female will not be so lustrous.

R.I.R. Club judges are instructed to pass "over-prepared" birds without comment, and to adhere strictly to the colour-standard, which calls for rich, brilliant red. I have not observed that this instruction has had any real effect on top-colour.

Few special breeding instructions are necessary. The breed Standard is such that double-mating is not required, so concentration should be on type and build, but ensure that hens have dark secondary markings on the wings. Mate up only birds with the desired long, brick-shaped body. Avoid cocks with cutaway fronts, or long sickles projecting far beyond the main tail, or defective colouring such as ticked hackle; and don't breed from females with short backs, rising saddles and quick lift of tail.

A tip for the beginner is that concentration on the long keel demanded by the standard is the only way of achieving brick-shaped bodies. Without this long breast-bone the bottom of the "brick" cannot be in keeping with the top.

One other point requiring serious attention is that development of rich colour and super-gloss or lustre frequently results in faulty, silky or frizzled plumage, "waxy" quill on back, and defective wing feather. Guard rigidly against these faults, which are hereditary.

There is a bantam club catering for the miniatures, but the large breed club also fosters them. They do not agree on weights, but otherwise adopt practically the same standards.

Bantam Club weights—Male 24–28 oz.: Female 20–24 oz.

Large breed Club weights—Male 28–32 oz.: Female 24–28 oz.

Serious defects include feather or down on shanks, or indications of plucking same: lopped comb or side sprigs, wall eyes; white showing in outer plumage, lobes more than half white; shanks and feet other than yellow or red-horn, or any deformity.

(To these defects it would be well to add Leghorn-type tails and the frizzled or Silkie-type feather often developed through concentration on lustrous dark plumage.)

SUSSEX

Sussex bantams are popular. The Speckled variety was the first to appear. Lights arrived later, and Buffs have been produced. Other colours are Brown, Red, Silver and White.

Main breed characteristics include a long body with wide shoulders, broad flat back, and broad square breast carried well forward, the long straight breast-bone or keel combining to form the typical Sussex outline. Wings are close, and the moderate-sized tail is carried at an angle of about 45 degrees from the straight back.

Carriage is vigorous and well-balanced, stressing the length of back. The neck is moderate in length and full-hackled. Eyes are prominent, beak short and strong, comb single, upright, and evenly serrated, close-fitting and moderate in size.

Plumage should be close and free from unnecessary fluff; and the hen's characteristics are similar to the cock, allowing for sexual differences.

Beak, in all colours, is white or horn, eyes red (orange in Lights). Comb, face, wattles and earlobes are red; legs, feet, flesh and skin white in all colours.

The Brown variety male has head and neck hackles rich dark mahogany striped with black and the saddle, back and wing bows are also the same main colour. Breast, tail and thighs should be black. The *female* should be dark brown with breast a pale wheaten brown; hackle should be striped with black and the body peppered with black.

The Red variety is rich dark red for both sexes with black stripes in hackle.

Silvers are very attractive. The hackle, back and wing bow should be silver with black stripes in the hackle. Tail and breast black with the breast feathers silver laced. The female is similar except that she does not have such denseness of silver in hackle; rather the neck and breast feathers are laced.

In the Light variety plumage is pure white, with black-striped neck hackle, the solid black centre free from shaftiness or false colour, and entirely surrounded by a white margin, without black fringing or black tips. Wings are white with black flight-markings, tail and tail coverts black. Undercolour is not standardised but is desired to be white for show; smut under saddle or back being a defect. This colour continues to be the most popular, but some fanciers have concentrated on the perfect hackle to the neglect of other qualities.

Figure 43.   Light Sussex Male. A Typical Prizewinner from Mr. J. Brannan

90

The White should have pure white plumage throughout.

The Buff should be of one rich even golden-buff ground colour, with green-black striping and markings as described for Lights. Buff under-colour is desirable, but dark colour is not penalised.

The Speckled cock has head and neck-hackle rich dark mahogany striped with black and tipped with white, and similar saddle-hackle. Wing primaries are white, brown and black— white not to predominate. Main tail black, white splashes permitted but not desirable; sickles black with white tips.

Remainder of plumage rich dark mahogany, each feather tipped with a small white spot separated from the ground colour by a narrow glossy black bar. Mahogany ground colour must be free from peppering, and underfluff should be slate-colour and red, with a minimum of white. (*Note*—in practice white in undercolour is penalised.)

The Speckled hen has flights brown, black and white, tail black and brown with white tip. Remaining plumage rich dark mahogany, each feather with a white tip and narrow black bar as described for the cock.

Breeding presents no real problems other than the necessity for concentration on type and character. Double-mating is not practised, the breed standards being so framed as to make it unnecessary.

In mating a pen of Lights, choose a male with good hackles and clean black tail. He may have smutty undercolour, but the hens should be clear; and if hackles are sound in both sexes a clear undercolour is satisfactory for breeding.

Maximum weights are: Male 40 oz.: Female 28 oz. These could well be reduced, having regard to the breed's undoubted progress.

Serious defects include—other than four toes; feather on shanks or toes; comb other than single, wry tail or other deformity.

WYANDOTTES

With its multitude of varieties and colours the Wyandotte can lay claim to being the premier soft-feather breed.

In self-colours there are Whites, Blacks, Blues, and Buffs; in particolours there are Columbians, Partridges, Silver-pencils, Buff Columbians, Black Spangles or Mottles, Duckwings, Cuckoos, Creles and Piles; and in laced varieties there are Golds, Silvers, Violettes, Blue-laced and Buff-laced or Chamois. All these colours have been shown, but not all are standardised varieties that would reproduce their like.

It is not easy to describe shape. The Wyandotte is a **bird of curves,** and should show no straight lines in its body from any angle. It must stand firmly on its own sturdy feet, with an appearance of greater depth than length, and a short U-shaped back. Its frontal and rear curves should be equal, and it should show very little thigh. The curve from throat to keel should match the rear sweep from back of keel to tail.

The body must be well balanced, the back and saddle broad, rising to the tail, which should not be carried very much lower than the head. This means that a Wyandotte, though upstanding, should not rear its head, but should have moderately low frontal carriage. Don't however, look for Cochin carriage or ball-cushions, and discard any birds without rapid rise of tail.

Wyandotte feather should be extremely abundant but not long; and although there must be plenty of fluff, this also should not be long, or you will get extremely baggy thighs and all the ills that go therewith. Breed always for soft tails—a Wyandotte cock is better with his main tail feather nearly as soft as his sickles and furnishings, all of which should be plentiful but short; and remember the tail in both sexes should be wide and well-filled, not flat and folded.

Figure 44.   Mr. J. Hey's White Wyandotte Cockerel. Winner of Many Top Awards

Other main breed characteristics are a neat helmet-shaped rose comb, firmly set, square and low in front, tapering evenly towards the back and finishing with a well-defined leader following the line of neck; top of comb covered with work (small rounded points) and all outlines convex and conforming to shape of skull. Earlobes are oblong, and wattles medium length—though in bantams wattles and lobes are too often circular, as in Rosecombs.

SHORT, DEEP BODY

The body simply *must* be short and deep, with four toes well spread, the alert and graceful carriage resembling the Brahma rather than the Cochin; and the female follows the general description of the male, allowing for sex.

Figure 45.   White Wyandotte Female. A Prizewinner for Mr. J. Hey

The beak of the Wyandottes should be bright yellow, but in marked and laced varieties may be shaded horn and yellow. (Yellow beaks are unobtainable in black males with sound undercolour, though standardised; so be content with black

93

shaded with yellow.) Eyes are bright bay in all varieties; comb, face, wattles and earlobes bright red. Legs and feet bright yellow (in Partridges they may be slightly shaded with horn).

Plumage of Whites should be pure white throughout, without straw tinge.

In Blacks the plumage should all be sound black with rich beetle-green sheen, the undercolour dark—as dense a black as possible. (This requirement means double-mating.)

Blues are an even blue throughout, medium shade preferred, free from mealiness, peppering or bronze and without lacing.

Buffs are sound even pale buff, with greater lustre on wing bows and hackle in males—otherwise perfectly uniform.

Partridges follow the usual black-red colouring and pattern-marking in males, with orange neck hackle and saddle-hackle shading off to pale lemon colour at ends; the hackle solidly striped with black, free from shaftiness or foul colouring, without black tipping, ticking, or fringing. The striping must not run through at tips.

The sound glossy black breast is free from red or grey ticking, and all undercolour black or dark grey, free from white (this is not easily attainable if pale lemon hackles are desired). Back and shoulders are rich bright scarlet, without maroon or

Figure 46. Mr. R. Nottrodt's Prizewinning Patridge Wyandotte Pullet

purple tinge. Tail, sickles and furnishings black with rich green sheen, showing no white at roots (again almost an impossible requirement allied with pale lemon hackles).

Wing bars should be glossy black, primaries solid black, free from white or grey; secondaries rich bay outer web, black inner web and end—the bay only to show when closed.

The female head and neck are rich golden yellow, the lower feathers finely-pencilled; remaining plumage a soft partridge-brown ground colour (often described as the colour of a dead oak leaf) covered with sharply-defined concentric rings of black pencilling following the outline of the feather; pencilling fine and even, clearly defined and with three or more distinct rings.

Red or yellow tinges in ground colour are objectionable, and shaftiness is a defect—especially on breast. Markings should continue well down the feather, and fluff would match body-colour. Short thumb-nail pencilling is objectionable.

Secondaries are brown on outer web, pencilled with black, and black on inner web—pencilling to show when wing is closed. Markings should be particularly good on saddle and breast.

The *Silver-pencilled* variety is the silver counterpart of the Partridge, being silver where the Partridge is gold.

Columbians have "Ermine" markings; pearl-white plumage with black-pointed tail, neck and wings, marked similarly to the Light Sussex (which see) but undercolour may be slate, blue-white or white.

Golds and Silvers (gold-laced and silver-laced) have sharp, crisp black lacing round edges of feathers, with rich golden-bay and silvery-white ground colour respectively.

Other colours are too numerous to describe here, but mainly their names explain themselves.

In addition to the breeding problems detailed earlier, there are several major difficulties in various colours and sub-varieties which make it imperative in some cases to use double-mating. Some colours, in fact, cannot be bred otherwise. Particulars of the methods employed are given in the special section on breeding for colour and markings. They need not therefore be detailed here, but the following notes will help.

Whites, Blues, and Buffs may be bred by single-mating, also Columbians and Buff-Columbians.

Blacks, which are required to have yellow legs with black undercolour in both sexes, *must* be double-mated. It is quite impossible to produce both sexes of show quality from one mating. So also with Partridges, which are probably the most impossible of all sub-varieties to produce on single-mating lines. This is because males and females are really two quite distinct varieties, not one.

Figure 47. Partridge Wyandotte Cockerel. A many times winner for Mr. P. Parris

Silver-pencils follow the same lines as Partridges, and must be double-mated if it is desired to breed them to published standards. Gold-laced and Silver-laced can only be bred accurately to British standards by double-mating—though American standards manage to avoid it.

The Standard for Buff-laced (white lacing on buff ground) should be attainable without double-mating; and the various violet-laced and blue-laced colours are usually (or at least frequently) all bred from one mating. It is not easy to understand the genetic principles involved.

Readers are advised to study the chapters on double-mating generally, and to apply the information there given to their chosen breeds.

The standards of the Wyandotte Bantam Club give weights as—Male 24–28 oz. : Female 20–24 oz. Although good specimens have frequently been bred at these weights, it is possible that slight increases might be advantageous.

Serious defects include—feathers on shanks or feet, permanent white or yellow in earlobes covering more than one-third of the surface; comb other than rose; comb flopping and obstructing the sight; shanks other than yellow; eyes not matching or other than bright bay; slipped wings (which should be penalised strongly); any deformity; conspicuous peppering on ground colour of laced varieties; double lacing in laced varieties.

MARANS

This breed originated in France. In the large breed it is noted for its dark brown eggs which are now being obtained from the bantams although there are the difficulties mentioned in earlier chapters.

They are compact birds with medium length bodies which are of good depth and width. The head is refined with a deep beak of medium size; eyes are large and prominent. Neck should be of medium length.

The shanks should be without feathers, white in colour, and of medium length, with four toes well spread.

Figure 48. Mr. Robin McEwan's Prizewinning Dark Cuckoo Marans Cock

Plumage should silky to touch and compact—not too much fluff. The bird should handle well being firm and well fleshed. The point to remember is that it is a table bird as well as a brown egg layer.

97

Figure 49. Marans Hen. A Winner for Mr. Robin McEwan.

Colours: *Dark Cuckoo*—each feather barred across with bands of blue black.

Golden Cuckoo—hackles bluish grey with black bars and pale golden shading on the upper part. Rest a mixture of bluish grey and black with rich bright golden and black bars across back and shoulders. Tail dark grey. Cuckoo markings throughout.

Silver Cuckoo—similar to Dark Cuckoo except a much lighter shade with mainly white in neck and upper part of breast.

Recommended weights are—Cock 32 ozs.; Cockerel 28 ozs.; Hen 28 ozs.; Pullet 24 ozs.

In judging 60 points are given to carriage, size and quality as a utility bird. Serious defects include coarseness or superfineness, lethargic in character, and feathered shanks. Defects such as a crooked breast bone, alien crossing or other evidence of foreign blood would call for being passed over by a judge.

CHAPTER 7

OTHER VARIETIES

THE RARE BREEDS SOCIETY has done a tremendous amount of work to revive breeds which were virtually extinct. The principal of these are as follows:

BOOTED: Once very popular here, but now more or less merged in Belgian d'Uccle, to which they were somewhat similar, though non-bearded. Known in Belgium as Sabelpoot.

CAMPINES: Two distinct colours—gold and silver. They have gold or silver hackles with the body features barred with the same colour.

CREVE-COEURS: Crested and horn-combed. Once bantamised but not seen for many years until recently. Were overshadowed by the more popular Polish.

CROAD LANGSHANS: Occasionally exhibited, but are usually first crosses between Modern Game and Pekins, as is evident when pads are inspected. They would be nice if the deep brown egg could be maintained.

DORKINGS: Occasionally seen before the war but never became popular and now are almost extinct.

IXWORTH: These are white and resemble Indian Game although softer in feather.

KRAIENKOPPE: Included in the *British Standards* in 1974 they look like becoming popular, but it is too early to be positive. Colours—Silver and Golden (see photograph).

LAKENFELDERS: Very lovely with their Mediterranean character, solid black necks and tails, and white body-plumage. Now very rare.

NAKED NECKS: Seen sometimes on the Continent but seldom, if ever, at shows in Britain.

NANKINS: Once the most wide-spread of all bantams and progenitors of nearly all buff varieties. Still standardised on the Continent but rare here. Resembles the Rosecap in type. The predominant colour is buff.

REDCAPS: Similar to O.E.G. in build and character, but with immense rose combs—hence their name. Seldom seen for many years past.

RUMPLESS (or Rumpies): Tail-less bantams, usually of Game character. Still seen occasionally but are not a distinct variety.

SCOTS DUMPIES: Were very quaint with their extremely short legs and long bodies, like very low-built Dorkings. Now only exist in the large breed, and those very rare.

Figure 50. Pair of Kraienkoppe Prizewinners for Mr. E. C. Ellis

SPANISH: Similar in type and plumage to Minorcas, but with the whole face white, and pendulous white lobes hanging below the wattles. The big breed is barely alive, but the bantams are very rare. Worth reviving and would be extremely interesting.

SULTANS: Crested, feather-legged, blue-legged and white in plumage. Now very rare.

SUMATRA GAME: Long in build and tail, with extremely rich green sheen on black plumage.

YOKOHAMAS AND PHOENIX: The famous long-tailed Japanese breeds, their chief features being an extraordinary development of tail and furnishings, which grew many feet in length and usually did not moult.

100

Bantam ducks and bantam geese have also been produced. They do not, I think, interest the normal bantam-keeper, hence are not described herein.

Figure 51. Mr. H. Parkinson's Pair of Croad Langshans. Magnificent Birds

Figure 52. Mr. T. F. Bower's CreveCoeur.

Figure 53.   A Rumpless Male

Figure 54.   Pair of White Yokohamas

Figure 55. Pair of Nankins bred by Mr. T. F. Bower

Finally, in case novice readers may query my not referring to Silkies, these are not classified as bantams by the Poultry Club. They must therefore be shown in the large breed section.

CHAPTER 8

HOUSING AND APPLIANCES

SELECT HOUSING CAREFULLY

ALL YOUR CARE in hatching, rearing and feeding will be wasted if accommodation is not equally well considered, so don't imagine that any old house will do. Apart from healthy conditions, different breeds need different housing, just as varying surroundings and locations demand different treatment.

Accommodation for Old English Game (which will roost in trees, if permitted, the year round) may be Spartan in simplicity, but would be totally unsuited to crested, feather-legged or inactive breeds.

If you are sufficiently good at carpentry to make your own appliances this is recommended. You may not save much money (in normal times the saving would be small); but you will be able to make just the type of house that suits your breed and your space. Where, however, stock houses are suitable there is little point in making them.

Let your houses have plenty of ingress for pure air, with outlet ventilators well above the birds when perching. Ventilation should never be entirely closed even on coldest nights; but don't make houses from any old scrap. Draughty buildings mean colds and disease. If your house *must* be cheaply made from odds and ends, cover all cracks and holes to exclude draughts and rain.

Lack of draughts, though, doesn't mean lack of ventilation. Give plenty of fresh air, because stuffy houses mean severe contrasts in atmosphere when birds leave the roost on a sharp winter morning.

Removable glass shutters over netted openings will give light and controlled ventilation; but don't close them except in really severe weather. The best lighting, however, is from small panels of vita-glass set near the roof apex. Top lighting is many times more efficient than wall-lights.

BACK GARDEN HOUSING

In back gardens compact non-portable houses and covered runs will be used, very different from requirements in fields or paddocks, where appliances may be constantly moved.

Back-garden fanciers comprise by far the greatest number of bantam breeders. Usually they are best suited by a small breeding house with covered run or scratching-shed attached, and a small open run for use in good weather. For most bantams a shed of around 4ft. × 6ft. will be adequate for a cock and three females. However, if a larger shed can be made available this gives more space for scratching.

Whether your housing be in garden or paddock, it is recommended that a covered run or scratching-shed should be provided. An alternative is a considerably larger house of the intensive type. This should be extra well lighted and ventilated, and a floor space 4½ft. square would accommodate about eight birds of the smaller breeds, or six of bigger utility character.

Even those restricted to a mere back yard need not despair. Many fanciers have successfully used double-decker housing and similar methods to overcome space difficulties, and have bred, hatched and reared chicks to maturity within a walled yard. So long as there is plenty of direct daylight failure need not be anticipated.

Double-decker housing, with fronts mainly of glass shutters over wire netting, can be adapted with the upper deck as an intensive rearing plant; the lower deck accommodating the breeding pen—which is, of course, provided with an open run.

GRASS RUN HOUSING

If you have ample space you naturally will not adopt back-yard methods, but will use field housing and grass runs.

Contrary to popular belief a grass run house should be of generous proportions. In inclement weather or when foxes or other predators are a danger, birds have to be locked up each night and, therefore, there should be room for them to scratch.

Probably no birds are more suited to the "folding system" than bantams, for which fold houses need only be small and therefore easily movable. They can be light and strong together, which is not easy in large units for big birds.

There are many standard designs on the market, in metal and wood, and I recommend a unit of semi-apex type about 8ft. long by 3ft. wide. If of purely apex design without sides, it should be a little wider; but combined houses and runs of this size will hold more bantams than any breeder is likely to run together and make daily moving unnecessary. If you keep not more than five or six birds in such an outfit, moving every two or three days will be satisfactory.

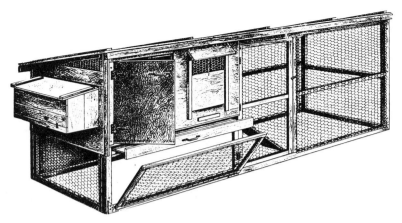

Figure 56. Space-saving House and Run Combined
(*Courtesy:* Park Lines and Co.)

THE BREED AFFECTS SUITABILITY

As formerly explained, houses and appliances used for Old English Game will differ from those most suited to less active varieties. For the former much less shelter is required; they are very active and hardy, and have no complications such as crests, feathered-legs, or trailing wings.

For Pekins and Japanese droppings boards should be close to the floor and perches set low; while for Polish or Polands droppings boards may well be omitted and slatted perches substituted close to the floor. Always consider such points before making or buying houses, because the accommodation most favoured for Rosecombs would prove quite unsuitable if your wandering fancy later turned to Cochins or Pekins, or Indian Game.

Rosecombs are a semi-hothouse breed, whereas Pekins (in spite of their feathered legs) are much more hardy and easy to rear on more open sites.

Bantam breeders nearly all have restricted space, and therefore need to make their appliances serve several uses. If rearing coops can later be adapted for cockerel boxes great economy is achieved in cost and storage room.

For baby bantams, runs with wire-netted tops open to rain and sun are death-traps. Rearing coops should either be of two compartment type, or be fitted with covered runs. Chicks then don't get alternately soaked in thunderstorms and baked in heat-waves. If chicks are to be reared on grass the runs should be made vermin-proof including wire netting on the base of the runs.

106

If you seriously contemplate showing your birds you cannot do without a penning room, or at least accommodation to train them for show. It is always best where possible to have a specially-arranged building which can combine food store and conditioning pen amongst its uses. An intensive house will do, or an existing shed can be adapted (e.g. an 8ft. × 6ft. garden shed). Show pens, made up of galvanised wire can be purchased, so with suitable shelving the conversion is straightforward.

ALTERNATIVE TO GRASS

Where open runs cannot be of grass they may well be gravelled, beaten hard and spread with sand, which may be cleaned by raking to remove droppings every week or so. They are best on a slight slope to run off water; and 18in. of boarding round the runs will protect against bitter winds and summer sun as well as prevent fighting.

Dustbaths should be provided—more for the hen's enjoyment than efficiency. Modern vermin-destroying perch paints and insecticides make dustbaths seem antiquated for their former purpose; and don't use ashes in them. Use fine dry earth and so avoid a lot of scaly-legs.

Figure 57. House and Separate Run: House 3ft. x 3ft. x 3ft. 3in. ideal for bantams. (*Courtesy:* Harry Hebditch Liimted)

Figure 57 (cont.)

Clean white sawdust is probably the best all-purpose litter. It is a deodorant and is vermin-preventing. It can also be sifted to remove droppings, and re-used; but don't use sawdust from sappy native timber or hardwood, or you will stain light plumage.

Some people say baby chicks will eat sawdust and become cropbound, mistaking it for cut corn; but are chicks really so stupid? A chick that didn't know food from sawdust would die anyway.

Sharp sand makes excellent floor-covering, and straw chaff is good, but peat moss is often dusty and dirty and plays havoc with the appearance of white birds and white-washed houses—though of course it is excellent in other respects, and is the "gardener's friend", especially in heavy soils.

If rats are prevalent you must either raise your houses about 9in. or 12in. above ground, to allow access for cats and dogs, or exclude them by tacking fine-mesh netting to the house walls, running the netting about 9in. into the ground with the lower edge turned outwards.

To exclude rats and sparrows netting should not be coarser than one-inch mesh; and if mice must be guarded against even half-inch mesh is hardly fine enough.

CLEANLINESS IMPORTANT

Clean houses regularly. Many people creosote them internally, but I prefer limewashing every year, with carbolic disinfectant added—it gives a much brighter house. The darkness of an internally creosoted house can, however, be partly overcome by using two parts paraffin to one part creosote.

A stout broom handle makes a good perch, but probably a rectangular one with the angles rounded off is better. For many reasons I think the best perch consists of three or four 1in. slats about ¾in. apart. This is particularly useful for encouraging young birds to perch. It can be set on the floor at first and raised a little each night.

CHAPTER 9

HATCHING AND REARING

IT USED STOUTLY to be maintained that bantams could not be hatched and reared successfully by artificial methods. This may have been true in those bad old days when the rule was "the smaller the better", but not now. Today we realise that the tiniest specimens are usually weeds, and we don't encourage them. The result is that our birds are quite capable of reproduction by any methods used for large breeds, provided due care is observed.

Quite naturally, the smaller and more highly inbred the variety the greater the care necessary; but many of our larger and more modern bantams are of a size and stamina that would have shocked old hands; and although they can't be produced in thousands, like commercial stock, they still present few difficulties to a knowledgeable back-yarder.

USING A SMALL INCUBATOR

There are two or three small incubators on the market which will produce satisfactory results with bantam eggs. The "visual" type is quite popular, possibly the two best known being the "Curfew" type and the VISION (Reliable) incubator. They have a perspex or plastic top through which the eggs can be observed.

The instructions issued with the particular incubator should be followed closely. Experience suggests the following hints should be followed:
1. Use fresh eggs not more than 7 days old. Mark them with the date in pencil, but use a felt pen to show the date when eggs are put into the incubator.
2. Eggs should be a normal, oval shape with strong, even shells (porous shells are not usually hatchable).
3. Locate the incubator in a room which is well insulated and free from sunshine or draughts (fluctuations in room temperatures are better avoided).
4. Keep the water tray topped up so that the eggs have adequate moisture.

5. Turn the eggs at regular intervals—usually twice per day.
6. When chicks start to hatch try to avoid opening the incubator.
7. Check the thermometer at regular intervals so that the incubator keeps steady at around 102°F.
8. Load the machine with a reasonable number of eggs; e.g. at least 20, or so few chicks will be hatched that rearing will be uneconomical.

Success with an incubator requires care, and application of the rules issued with the machine. Checking for fertility is advisable at 7 and 14 days. Any clear eggs should be removed. With experience it becomes possible to use a strong torch and detect fertile eggs which have become "addled". However, in the early stages the beginner is advised to remove the "clears" only.

Once the chicks have hatched they can be transferred to a broody hen—one which has been sitting for at least two weeks. Alternatively, a brooder can be used; this may simply be an infra-red lamp placed in a suitable shed.

In hatching under hens you need not confine your attention solely to bantams as broodies; but it is not wise to use heavy hens of 7 or 8 lbs. weight, or you will suffer broken eggs and crushed chicks. If you must use large hens, try medium-weight utility birds such as Buff Rocks, which sit lightly on the eggs and are careful mothers.

Smaller broodies are, of course, better, and usually the bantam breeder gets enough sitters amongst his own flock; but those who require stocks of broody hens might consider keeping a small flock of Silkie-crosses. These are better than pure Silkies, which have several disadvantages. They desire to lay and sit again somewhat quickly, whereas bantam chicks require an extended rearing period; they are very susceptible to scaly-leg, which is transmitted to chicks; and their silky plumage sometimes wraps itself round the necks of chicks and strangles them.

The best broody hen you can get is a Silkie crossed with a clean-legged variety such as the Wyandotte. This is a great favourite with large-breed fanciers and has even been produced commercially. Their "staying-power" greatly exceeds that of pure Silkies.

Whatever you do, avoid scaly-legs in broodies, or it is absolutely certain to attack your chicks. Even hens that have been treated are not good—the paraffin or other cure makes them restless, and they sometimes quit before hatching.

Your hen is best not heavier than about 3 lbs., and is therefore usually a bantam cross or pure breed. Dust her well with insect powder—or alternatively, when trying her out, put a spot of nicotine sulphate perch paint on each dummy egg. The fumes generated will kill every insect on her body, and the use of perch paint avoids the handling necessary when applying insect powder. A roughly-handled broody gets nervous and may not readily settle down.

For broody hens cheap nests like apple boxes have much to offer, which can be burnt after use and so avoid vermin. Fresh nests for every batch of hens are excellent.

It is often recommended that the foundation of the nest should be earth, or a turf moulded saucer shape. The chief virtue of this is that it shapes the nest nicely; not (as most writers insist) that it provides moisture for the eggs.

NEST OF SOFT STRAW

Whether or not you use earth as a base, the nest itself should be of soft straw. Hay is not so good, as it is dustier and more likely to attract vermin. Oat straw is best; make the nest saucer-shaped, put a few crock eggs in it, and show it to the hen. More often than not she will walk on and settle down—though if she is a stranger she may need coaxing and shutting on the nest. Some of the best are restless for the first few days.

When she returns to her nest of her own accord after about ten minutes' feeding and watering she may be trusted with the eggs she is to hatch. Give her these at night, when she is quiet and settled. Don't give her every egg she can cover, particularly in cold weather. Two less are better than one too many, or chilled eggs and a bad hatch will result.

Don't startle hens by sudden noises. Keep them quiet or you will experience trouble. Their vicinity is no place for hammering up new sitting boxes or repairing wooden fences. Let them have ten or fifteen minutes daily off the nest and see that their drinking water is clean. They are best fed on grain if possible, as soft food is likely to promote diarrhœa.

Nests fouled by broken eggs or excreta should be cleaned, the straw replaced, and eggs washed in warm water. This shouldn't occur if the hen is comfortable, but it happens all the same. If the hen has diarrhœa, slip down her gullet a lump or two of gum arabic moistened in cod-liver-oil twice a day for two days. This usually effects a cure. If not, try feeding with bread soaked in milk, squeezed dry and dusted heavily with french chalk.

At four or five days, test fertility by holding eggs at night before a strong light. The best way is a box with a hole for the egg and an electric light inside. Fertile eggs will clearly show blood veins and the germ spot, the whole at the fourth day looking like a spider's web. With light shelled eggs you can test readily at the third day, but with dark brown eggs the fifth day is usually best.

When testing eggs (particularly in the early season, when infertility is frequent) you will see the advantages of setting several hens at once. The fertile eggs from four hens will usually be sufficient for three, and the extra hen can be given a fresh clutch.

Bantam chicks, by virtue of their tiny size, often have a tough job during hatching; therefore when utility breeders say (quite rightly) that chicks unable to hatch themselves without assistance aren't worth rearing, don't apply the rule too strictly to bantams.

Often highly-bred bantam chicks need helping out of the shell, particularly in the smaller breeds. Bear in mind that a Rosecomb chick at three weeks old is smaller than a day-old large breed, and you will understand our methods must differ from commercial practice.

There is often, in bantams, twenty hours or more between the first chipping of the egg and the hatch, so don't be in too great a hurry to help chicks out. All the while they are strongly tapping and chirping, don't interfere unless progress ceases. Then very gently break away the shell round the circumference where already chipped, stopping at once in cases where veins in the membrane covering the chick are full of blood. Only when these are practically dried out should the membrane be torn away; then, gripping the chick's beak beween finger and thumb, help its head free and put it back under the hen. The chick will do the rest.

Chicks should hatch on the twentieth day—a little earlier than big breeds. When the first chipped egg is seen, do not allow the hen off the nest till the hatch is over. If necessary she can be fed and watered on the nest.

Bantam eggs usually hatch better if they get a little extra moisture. Under hens this can be arranged simply by damping the breast feather (what little of it remains during incubation) or by sprinkling the eggs with warm water two or three times from the seventh day to the seventeenth. Wiping the eggs with a warm damp cloth is also helpful.

WHEN CHICKS HATCH

Leave chicks under the hen for about twelve hours, but give her food and water on the nest. The next day, leave them in the nest while giving the hen a good feed of grain and a drink in the coop she is to occupy. Then give her the chicks and she will settle down. For almost the whole of the first day she will keep them under her wing, so let her alone. The second day scatter chick crumbs to encourage the babies to move about. They will investigate curiously, especially when the hen clucks for them to eat.

Give water from the commencement, in a shallow trough which the hen cannot soil. No harm comes from clean water, no matter what old hands say. It is stagnant, foul water that causes trouble. You wouldn't give a tiny child dirty water to drink, so use common sense.

Many people used to advocate hard boiled egg and bread crumbs for the first few days, but I disagree. Hard boiled egg is not only somewhat indigestible, but is troublesome. A little biscuit meal scalded with milk and dried off with baby chick mash is better; and probably the most successful method of rearing is to feed completely dry for the first three weeks on baby chick crumbs in hoppers alternated with chick grain. If you use soft foods feed them nearly dry and merely warm, not hot.

Feed frequently—say every two hours at first, gradually reducing the number of meals until at a month old they are only fed four times a day at most; but the dry feeding method outlined above is a great time-saver, because the baby chick crumbs are constantly in hoppers before them.

As an appetiser, after the first couple of weeks the first and last meals each day can be soft food, and this gives an excellent chance gradually to wean the chicks to growers' mash and more normal, less expensive foods. Don't spare expense, though, in baby chick food. They eat so little the first three weeks that the heaviest economies present practically no saving.

A little chopped grass is excellent for chicks reared intensively. Cut it very short with scissors and they will eat it greedily without danger. I suppose most baby bantams are reared intensively for the first few weeks. Once they are outdoors they can run on grass or be given a turf on which to scratch. A good way is to put a large turf in their run, scatter their grain food on it, and see them scratch.

Cooked chopped meat (in small quantities) is good. Feed regularly and well. Never *feed* for small size: *Breed* for it instead. The old days when bantam experts recommended feeding on roasted wheat and other devitalised foods are happily past and gone.

INTENSIVE REARING

You will gather that if you haven't much outdoor space you needn't worry. Bantam chicks can be reared intensively with complete success, no matter what experts may say to the contrary. They must, however, have ample direct daylight (not through ordinary glass) and about one per cent of cod liver oil in their food. This is best stirred into the chick grain just before use.

Greater success attends indoor rearing than outer runs, particularly for the first three weeks. Common sense in feeding and management must be your guide.

Leave your chicks with the hen as long as she will care for them. They are seldom ready to leave her until they are eight weeks old or more; and when they leave the hen separate the sexes. They will thrive much better.

It is a good thing, when cockerels get big enough to fight, to run an adult cock with them. He will act as policeman and keep the youngsters in order.

113

Even for breeding, stock bantams are hatched later in the season than large birds. For health and development, and for future breeders, there is nothing like May hatching; while for show (and for small size) later will suit. Many show birds are hatched as late as July, but this is not recommended, and in any event doesn't apply to every breed.

April hatching is good for birds with flowing feather, like Rosecombs; and in Modern Game later hatched birds do not acquire the reach and lift of early chicks. The earlier chicks are hatched, within reason, the more rapidly they mature and the better they thrive, though they also get too coarse in build.

Probably it would be correct to say that the ideal hatching season is from April to June inclusive. Earlier chicks get too coarse and later chicks are often puny and troublesome. Nowadays, too, we don't carry on the stupid old practice of hatching in August and September to get tiny birds for show. If we hatch after June it is only because we have failed to get enough chicks earlier.

The old-time idea that artificial hatching and rearing of bantams was a failure has been proved wrong over many years. Chicks can be reared artificially with great success under an infra-red lamp.

However, for the best results "black" heat is better, thus allowing a lighting programme to be used which gives the chicks the optimum number of hours of light.

Artificially-reared chicks grow more quickly because they get more heat. Under a hen they only get warmth when she sits down to brood them. In a heated rearer they can constantly run to the heat after feeding, hence they get more continuous warmth.

In brooders, too, chicks feather more quickly, and they make an extra crop during the rearing stages—or alternatively get their adult feather earlier. This is sometimes an advantage, but not always.

Incubators score heavily over hens—they don't break eggs and they don't crush chicks. Similarly, brooders have two great assets—they have no feathers to harbour vermin and they don't scratch. At times it is a pitiful sight to see tiny bantam chicks battered to death through the misdirected energy of a scratching large-breed hen; and it is just as pitiful to see a baby bantam gasping out its life under the weight of a heavy hen too stupid to realise she is standing on something other than her own feet.

It is a good thing, if you rear artificially, to hatch a week or two later than under hens, to compensate for more rapid growth and quicker feathering, otherwise your chicks may turn out bigger and coarser than you expect; but don't be misled by advice from those who haven't tried it. If you find it difficult to get broody hens, try incubators and rearers. You won't regret it.

When rearing artificially use intensive methods in preference to outdoor brooders—though the latter can in the right circumstances, and with the right design, be extremely successful. It is true that chicks need not be allowed outdoors at all for the first couple of months: But don't deprive them of fresh air and direct daylight.

When you've finished rearing, don't wait until next season before disinfecting all appliances; and if you don't maintain your own breeding stock, don't expect to buy day-old bantam chicks. They are too tiny to travel safely, so buy breeding stock or eggs and hatch your own.

In stressing the value of overhead shelter for baby bantam chicks it is not my intention to decry open-topped runs so much as to emphasise that open runs should not be used except in conjunction with some form of shelter.

Rearing isn't finished when chicks leave mother hen or brooder. They still have to be brought to near maturity before they are fit for selection, and for segregation into breeding stock, show specimens and sale birds—the first-mentioned being by far the most important (though few novices can be persuaded to believe it).

A few special appliances are necessary for growing stock, but these can be of readily-adaptable type, suitable at different seasons of the year for use as cockerel-boxes, conditioning pens, or single-mating houses when breeding from pairs—which can be done when mating for special points.

Ready-made coops and sheds can be purchased from specialist suppliers. Unless the fancier is good at wood-work the cost of purchase will be worthwhile. A joiner-made bantam house, creosoted regularly, will last many years.

CHAPTER 10

FEEDING ADULT STOCK

FEEDING AFFECTS SIZE

EFFECTS OF FEEDING on size and feather should not be ignored. Plentiful soft cooked foods, worms and garden grubs will develop game chicks so big and feathery that at four months of age they are larger and coarser than adults, useless either for exhibition or breeding. Such a diet may be useful for big breeds where bulk and bone are required, but bantams need to be kept down in size.

The general principle is that hard feeding produces short, hard feather and vice versa. Game breeders will therefore use grain as their staple food, with only small amounts of soft mash; whereas breeders of soft-feathered varieties will feed on soft foods, with hard corn merely as a small last feed of the day.

In neither case should foods be of a fattening or frame-making nature; yet for the sake of health and stamina we must avoid malnutrition and debilitation.

Bantams lay eggs much more freely in warm weather than in winter; so if eggs are wanted early in spring, feed generously and add a little extra meat meal to their soft food.

Where birds have a good grass run they will need very little other green food, and if at liberty they will require less feeding than if confined. A small ration of soft mash in the morning, with grain at night, is a good method—though the procedure can readily be reversed if desired.

AMOUNT OF FOOD

One good handful of this soft food is plenty for three adult bantams for the morning meal—more or less depending on breed and size. At night, one good handful of grain (gripped palm downwards) is similarly enough for three birds. Don't exceed these amounts except in the cases of over-sized birds of utility type. Overfeeding produces multitudes of troubles.

With modern poultry foods it is possible to obtain layers' crumbs which can be kept in hoppers, thus cutting down on labour. The specially balanced food produces more eggs and better fertility.

Grit is, of course, important—flint for digestion and limestone or oyster-grit for egg shells.

Corn can be fed in the litter to encourage exercise; but if your grass runs get heavily tufted scatter the grain there and see the birds tear down the tufts.

Green food is best given daily in wire racks. These may be merely wire netting baskets fixed to outer runs, or specially-made racks.

Use as little maize as possible for white birds if you wish to keep them white; but for blacks it increases surface sheen—so does the occasional addition of small quantities of flowers of sulphur to the soft food. Don't use it with white breeds if you desire pure colour.

SPECIAL FEEDING

During the moult a little linseed stewed to a thick jelly and added to the mash aids feathering and condition. A tablespoonful of the jelly for every ten birds every other day is plenty. Sour milk is also an excellent addition to the mash.

When birds are penned up for show over extended periods, be sure to include cod-liver-oil in their diet, and occasional tit-bits of minced raw meat. Canary seed and hempseed are also, as previously mentioned, excellent conditioners, which are often necessary because show birds are frequently fed more heavily than breeding stock. Watch their condition carefully, and when birds under training seem lethargic and inactive a little liver pill often works wonders.

Some judgement is necessary—birds in training should be kept comparatively inactive without putting them out of condition. They must not be overfed, but on the other hand should not be so eager for food as to fly at the pen front when they see you approaching. Handling is the solution—birds frequently handled don't get out of sorts.

We are often told to scald feeding troughs regularly, but I don't like creating labour. In my own yard I keep a water butt into which I regularly drop all feeding appliances and water troughs. They clean up extra well after a day's soaking.

Clean water should be constantly before all birds. Keep it out of the sun, and in summer stand drinkers on the north side of the house in open runs.

Drinking fountains can be purchased which allow birds to drink easily and yet do not allow the water to be fouled. Remember that water is absolutely vital; birds cannot survive without it for very long.

CHAPTER 11

SPECIAL BREEDING PROBLEMS

THERE ARE MANY problems in poultry breeding common to all similar or similarly-marked varieties. It is therefore well to deal with them in a special chapter, and avoid having to repeat details under each breed section to which they apply.

DOUBLE-MATING*

Novices are frequently puzzled by the term "double-mating", and by reading of cockerel-breeding and pullet-breeding pens. They imagine from these terms that it is possible to mate pens which will produce only cockerels or pullets. This, of course, is not so.

The term "cockerel-breeding" is applied to a pen of birds so mated that they will produce extra good show-type cockerels; and the pullet-breeding pen can similarly be expected to produce better females than the normal. These special matings are used in cases where colours present difficult problems, or where markings are of such an intricate nature that they clash in the sexes, and are incompatible with one another.

The best example of double-mating for colour is the Black Wyandotte, in which (like the Black Leghorn and all other black breeds with yellow legs) the sound black undercolour demanded in males is not reconcilable with the possession of a rich yellow leg in both males and females.

Correctly (as in wild birds) the counterpart of a male with rich black plumage and bright yellow legs is a dull-coloured female with black or dark legs; and in demanding yellow legs in females, together with dark undercolour in males, we are looking for something unnatural, to achieve which we must make special matings.

Similarly, the best example of double-mating for markings and colour combined is the Partridge Wyandotte, in which males with solid black breasts and solidly-striped neck hackle are not compatible with the pencilled plumage demanded in females.

---

* This section on double-mating is included as originally written. However, the modern tendency is to avoid where possible.

This example is more than a mere divergence in colour schemes —males and females are really two different varieties. The male is a black-red, the female is gold-pencilled. To produce good standard specimens of both sexes two pens are absolutely essential.

The lack of easily-understood technical terms makes it difficult to discuss these problems without repetitive phrasing. Unfortunately the jargon of poultry and its literature have perpetuated false descriptions. Thus even the general term double-mating is incorrect, and should more properly be called *single-mating,* since each is a mating to produce a single result, not two.

When a standard insists on an unattainable purity of colour in both sexes, or on markings in one sex completely opposed genetically to markings in the other, obviously the Standards are wrongly framed and should be re-drafted.

Probably none of the old hands would agree with this—the custom in old days was deliberately to make things harder when breeding for show points; but Standards being what they are, until they are changed we must use matings that will produce birds which conform to them.

In describing methods, therefore, let us do so on general lines applicable to all breeds and colour-varieties. The problem of reconciling yellow legs and black plumage is practically the same in Wyandottes as in Leghorns; and difficulties in producing so-called Partridge males and pencilled females apply to Cochins or Pekins as well as Wyandottes, and in a limited degree also to Brown Leghorns. They would apply to Rocks if the Partridge variety were revived in that breed; and there is little point in repeating the same instructions over and over again.

YELLOW-LEGGED VARIETIES

The main varieties to which the problem of blacks breeds with yellow legs apply are Black Leghorns, Black Wyandottes, Black Rocks, Black Barnevelders, and Black Japanese; also in a minor degree, Black Frizzles and Black Cochins or Pekins.

Japanese or Pekins are seldom double-mated seriously, and Black Barnevelders are not often seen in bantams.

Sound undercolour in males, accompanied by rich yellow legs in females, can only be achieved by realising that show males and exhibition females are, to all intents and purposes, two distinct varieties. To produce show females you must use a sire with light or white undercolour; and to breed males with sound undercolour you must use dark-legged females.

The cockerel-breeding pen, then, consists of a first-class standard or show-type male, with rich black undercolour, mated to females with dark legs. The extent of this darkness varies from a mere dusky surfacing over the whole leg to a practically black shank; but in all cases the pads of the feet and the backs of the shanks must be yellow. Don't breed from females with all-black legs or with very dark brown eyes—both points are evidences of excess pigment that will probably result in black-legged cockerel progeny.

There should be sound plumage colour in both sire and dam, and plentiful green sheen. Don't be influenced into using sooty black females—the best way to produce green sheen is to use plenty of it.

Your ideal cockerel-breeding females, therefore, will have sound black plumage and undercolour, good type and head-points, good eye colour and shape, and dusky legs; varying in depth of black on shank, but always with a fair amount of yellow present. Mated to a standard male these will produce good cockerels, whose undercolour will be sound, and whose legs and feet, though possibly dark in chickenhood, will become clear yellow as they mature. (Cockerels with clear yellow legs in baby-hood usually have poor undercolour when mature).

The production of exhibition pullets requires opposite methods. To females of standard colouring and show character mate a male with very rich green sheen, rich black top colour and clear yellow legs, but with light undercolour.

QUESTION OF UNDERCOLOUR

The amount of light undercolour will depend upon his age and upon his mates. If a cockerel he will need very little white or light undercolour, provided he is definitely bred from a pullet strain; but pullet-breeding males become lighter under as they mature.

A cockerel would have a small amount of white in saddle and back, and be moderately light under his neck; whereas the same male, at two years old, would be almost completely white under throughout back saddle and neck, and possibly would also have some white in flights. Do your best, however, to find a male with sound black quills in his wings.

Another cause for variation in the extent of his light under-colour is depth of colour in the females with which he is mated. Some show females are of distinctly silvery undercolour after a few generations of pullet-breeding. This calls for two things—less light undercolour in the male, and possibly a later introduction of a little cock-breeding blood to provide more pigment.

For several years I mated over 20 pairs of birds each season in special experiments, and discovered many points too numerous to detail here. If you keep accurate records and toe-punch all your chicks you will discover lots of important but small items for yourself; and toe-punching will help you to avoid mixing strains. This must never be done except under complete control.

In Black Leghorns similar methods apply, but pullet-breeding males are often used with decidedly white sickles, in addition to 'white undercolour. Each breed dealt with provides its own problems, which only experience will solve.

You will learn that in cock-breeding some males will not clear their legs until the second season; and some cock-breeding hens become almost clear yellow in shanks when they are several years old. Make sure therefore that your pens are composed of birds correctly bred.

If you can't be certain of their breeding, you will have to mate up the best way you can, and produce your own cockerel and pullet strains. In doing so you will breed rather a mixed lot for a couple of generations; but after a few years you will find yourself with cock-breeders that can be depended upon to produce practically all males with yellow legs and sound or nearly sound undercolour.

Don't breed from birds with purple barring, and don't use birds of bad type merely because they have good colour. Even after rigid selections and culling, you will find that each year your best birds all come from about three females.

BREEDING FOR COLOUR AND MARKINGS

Nowadays we suffer from the fact that old-time fanciers had false ideas of breeding for colour and markings. They knew little or nothing of genetics; and their practical knowledge and rule-of-thumb methods were not accurate enough to teach them to refrain from "making the job harder".

Always they bred for and cherished that which was hardest to attain. They did not realise, for instance, that in so-called Partridge varieties, where they standardised fine concentric triple-pencilling in females, with black breasts, solid striping in necks, and lemon-coloured hackles in males, they were being so illogical as to accept two sub-varieties (Black-Red males and Gold-pencilled females) as one.

They were, in fact, attempting the impossible; and in their so-called double-mating they were really breeding two varieties. In cockerel-breeding they were not merely producing correct Partridge males, but also correct Partridge females; while from the pullet-breeding pens they were breeding both Gold-pencilled females and Gold-pencilled males.

In these "partridge" matings let us again deal with Wyandottes, though the general problems apply to any breed of approximately similar markings—that is, any breed in which the male is of partridge colouring and the female is gold-pencilled. First, however, let us realise the essential distinctions from the more correct partridge colouring of Old English Game, the near-partridge markings of Welsummers, and the less-definitely-segregated Brown Leghorn.

In O.E.G., where markings of Partridge females show fine stippling instead of concentric rings of pencilled markings, good birds of both sexes can be bred from one pen. The same remark applies to such breeds as Welsummers.

In Brown Leghorns, where a completely sound breast is demanded in males, with very fine delicate pencilling and absence of ruddiness in females, a modified form of double-mating is practised; but the problems are not very difficult, and good specimens of both sexes are frequently bred from one pen. In Partridge Wyandottes, however, this is impossible.

Even by double-mating it is not practicable fully to achieve standard requirements. You can never obtain, for instance (in the same cock) palest lemon hackles, absolutely sound black breast with dark undercolour, and a tail free from white at the roots. It can't be done—you must always compromise between pale lemon hackles and soundness of undercolour.

PULLETS OR COCKERELS?

If you are wise, in breeding these Partridges you will stick to one kind—pullets or cockerels; but if you *must* breed both, keep them rigidly separate, and never mix them in any circumstances.

To produce show cockerels, use a sire with particularly dense, solid black striping in neck and with broad gold fringe to each feather, shading off to pale lemon. Striping should neither run through at tips nor have pale shafts, and hackle should have no black tips or fringing. The head should appear clear orange at cap, shading down to pale lemon at base, with practically no black markings visible. Saddle hackles should be similar, but in bantams are seldom so pale in lemon fringes.

If this bird has bright scarlet back and shoulders, not shaded with purple or maroon, solid black primary feathers free from white, and rich bays on wings when closed, he will be good. We want also a sound black breast without ticking, but we shall certainly not find this bird with dark undercolour. We must put up with that, and look for as little white as possible at the root of the tail; and although we have been looking for a sound select the nearest we can.
unticked black breast, we must remember that a cock almost perfect in breast as a yearling is likely to show some red ticking later in life. We shan't, in fact, find a perfect cock, but we'll

In mating this cock we must compromise. To give him hens possessing sound concentric pencilling would be futile. Males so produced would have pencilled breasts and defective striping, amongst other things; and since we want solid striping and solid black breasts, we pick first a hen that will give us these points. She will be practically a brown-red, with very dark body showing a great amount of black, heavily peppered all over, with pale lemon-coloured hackles solidly striped with black. This mating will give density of black and solid striping in the male progeny.

As you won't get all you want from one hen, you will need one or two others of different character. A second hen could be similar in most respects, but paler in ground; rather darker than a show female and with much heavier, denser markings. She may perhaps show a tendency to concentric pencilling, but so long as this pencilling is very dense and broad it will not matter. Again she should have solidly-striped neck, and a little ruddiness on wing will be an advantage.

Once you have bred a few generations you will add other types to your cock-breeding hens; but remember that you are building up your own strain, and can't at first expect to breed all your cockerels dead alike. That will only come when you have fixed your strain and methods.

DON'T USE SPORTS

In concentrating on the production of pale lemon hackles you will possibly breed some white sports. Using these in the breeding pen will produce pale lemon hackles, but don't. You will spoil your strain, not improve it. You don't want to perpetuate the tendency to produce albinos.

The principles involved in pullet-breeding are in some ways the reverse of cock-breeding. Solid striping is not wanted in the sire's neck. Ticking and broken markings are better. Orange hackles are wanted, not pale lemon; and instead of sound black you want red pencilling or lacing on breast if obtainable—if not, plentiful red ticking instead. His back, shoulders and saddle should be broken in colour and markings; and if there is pencilling and ticking on wing bows, tail and thighs treasure him.

For pullet-breeding you want a male with as many female characteristics as possible, so make sure he is pullet-bred—though if he is as described he could hardly be anything else.

The mates for this cock would be hens of show character and colour. The darkest advisable ground colour is the soft brown of the show hen. Some of the hens might be a little paler, to counteract the darker pigmentation of the cock. Each feather on a good female will have three concentric pencillings of black, with an outer fringe of ground colour. Pullets often show part-barring, but this disappears with age.

While building up your strain put your best hens in the breeding pen—they're more valuable there than for show; and this means your breeding hens will have neck hackles pencilled or ticked, not striped, with well-pencilled breasts free from light shafts, and brown fluff to match ground colour.

BREEDING SILVER PENCILS

Silver pencilling is merely a silver edition of the so-called Partridge (which is really gold-pencilled). General methods resemble Partridge-breeding. The worst failings in Silver-pencilled males are lack of clear silver-white colour in hackles, and ticked necks. They are not often bred to high quality, but are frequently better in type than Partridges—possibly because they are not usually double-mated.

Methods of breeding would be more or less as described for Partridges. Thus a pullet-breeding cock would have a black breast laced with silver—the more lacing the better, perhaps with pencilling on thigh fluff, shoulders and wing; and cock-breeding females will be mossy or peppered (not pencilled) with striped hackles.

These remarks apply also to Dark Brahmas, as well as any other silver-pencilled variety. It is a curious criticism of the old-time fancy that in Brahmas the Silver-pencilled is called the Dark variety; and that what is known as barring in the Campine is described as pencilling in the Hamburgh.

There are plenty of other breeds so standardised as to need double-mating. The idea underlying all methods is simply to mate an exhibition-type male to females with male characteristics if you want to breed specially good cockerels; and for producing good pullets mate show type hens with a male possessing female characters or markings. This is particularly easy to study in Indian Game.

The male Indian is almost a self-black, only slightly broken with bay or chestnut, and with very rich green sheen. It naturally follows that to produce good sons he must be mated with hens of male character, with as much black as possible in the form of heavy lacing, not open but broad and dense. After breeding for a season or two you would produce cock-breeding females that were very heavy in black, almost single-laced, and nearly black on back.

Similarly, for breeding good pullets you would mate exhibition-type hens with a cock inclined towards female markings. In pullet strains cockerels are sometimes laced almost like females when young. These will be your future pullet-breeders.

*　　　*　　　*　　　*

A chapter like this could go on indefinitely detailing the various matings possible in nearly all breeds to improve the sexes, even where strict double-mating is not practised. Sebrights, for instance, were double-mated in the old days, but in a very simple form. Good show pullets were bred from a finely laced cockerel mated to heavily laced females; and the best cockerels from a heavily-marked male mated to finely-laced females.

Other laced breeds, such as Wyandottes, were bred by similar methods, modified as found necessary for each variety.

Nearly all Mediterranean breeds can be improved for show by a systematic course of double-mating for headpoints—to produce erect combs in males and lopped combs in females. A similar remark applies to Rosecombs, where cock-headed females will give better combs in male progeny. These simple special matings are, however, mentioned and dealt with under breed descriptions.

124

Perhaps things ought to have been so arranged in all breeds that double-mating was unnecessary; and doubtless if we were now drawing up new standards we should, in the light of present knowledge, so arrange it. Much of the competitive charm of breeding, however, would be lost; and it is certain that old die-hard fanciers would object strenuously.

Yet there seems logical ground for renaming such duplex varieties as Partridge Wyandottes, and for encouraging in them (and in breeds like Dark Brahmas) male plumage and markings that are compatible with female standards. Males would be just as beautiful as now—but perhaps good ones would be too easy to produce. Is that the answer?

CHAPTER 12

PREPARING FOR SHOW

TRAINING BIRDS

SHOW TRAINING is not necessarily show preparation. Apart from any necessary washing and cleaning, the birds must be schooled to show their points.

Take up untrained birds in the evening after going to roost and put them in show pens for the night. If you exhibit regularly you should have a penning room.

Leave the birds to settle down, then by artificial light stroke them quietly down back and throat, talking all the while. It is amazing how much they will endure by lamplight. When they will stand quietly in one position, take a small stick and go through it all again. If a bird carries its tail too high, stroke it down; if too low, lift it. A well-trained bird gets used to the judging stick and acts accordingly by lowering the tail the moment a stick is placed on its back.

Bad habits, such as low wing carriage, can be corrected by training. Touch birds under their wings, lift them up, and induce them to tighten loose carriage. When there is nothing radically wrong it is easy to school them out of careless habits; for example, by tapping legs or hocks to improve style and carriage.

Specialised training is easy, such as giving a piece of raw meat or other tit-bit to a Modern Game bantam high up in front of the pen, to encourage lift and reach; or similar methods of encouraging low-fronted carriage in Pekins, Orpingtons, etc. These methods are mainly used in daylight, after birds have had a couple of nights' training by artificial light. You can handle a bird more and more until he gets ready to show himself the minute he sees you.

Half of the preliminary training is unnecessary if your birds have been handled in chickenhood. The more young chicks (and their mother-hens) are handled the better. Some of them could then be successful at shows with very little pen-training.

While penned up, a little lean raw chopped meat, bread and milk, and special feeding like canary seed, will assist conditioning for show.

DUBBING GAME BIRDS

There are differences between game birds and soft-feather varieties when it comes to preparing for show; and as Game birds almost always come first in show classification, let us consider first one of the problems affecting Game cocks—namely dubbing.

It is customary to dub all male game birds (that is, to remove combs, wattles and earlobes) otherwise they stand no chance in the show pen. Much has been said for and against this practice, which however is not so cruel as many people pretend.

The problem really originated as a matter of expediency. Game cocks are extremely pugnacious, and dubbing was found necessary to prevent them tearing wattles and combs to pieces. The damage caused by a pitched battle is far greater and more painful than the operation of dubbing, which should not take more than three or four minutes. I have dubbed hundreds and never lost one; but I have lost many undubbed birds through fighting.

Get some dubbing scissors, which are curved and allow the comb to be cut to the desired shape. Have someone hold the bird firmly, with a leg and a wing in each hand, facing the operator.

The operator holds the bird's comb firmly with finger and thumb, removing first earlobes, then wattles; taking off the latter from back to front with one clean cut. Do this on both sides, then insert the left thumb into the beak, holding the lower mandible between thumb and forefinger. Stretching the neck full length, cut off the comb with one long clean cut from back of head to beak.

For Modern Game both comb and wattles are dubbed very close to the skull to give a clean-cut, snaky appearance; but for Old English Game sufficient of the comb is left to give a hardy, masculine style—a half-moon shape.

The head should then be bathed in cold water, which will stop most of the bleeding; and in good weather the bird can be returned to his run. In nearly all cases, as soon as you put him down he will start crowing, so the operation is obviously not as painful as sometimes pretended.

WASHING FOR SHOW

All light-coloured birds (not merely whites) are greatly improved by washing for show; and many a good bird has lost prizes for want of a wash, or through being badly dried.

The process is simple. Three bowls of water are needed, a soft nail brush, soap or washing-up liquid, and a plentiful supply of soft water. If your water is hard you must take steps to soften it, or alternatively use rain water.

Work at the kitchen sink if you can. The first bowl should contain water fairly hot, but not too hot for your hands. It should be deep enough to immerse the bird with head above water. Work up a good lather, stand the bird in the bowl, close his beak and nostrils while dipping his head, then scrub comb, face and wattles with the soft nail brush and soap, wiping off with the sponge.

After soaking, give the legs a vigorous scrubbing. Dirt under the scales can be removed with a match-stick sharpened to a point.

Holding the bird with head above water, open up the plumage and soak it thoroughly. Next work soap well into the feathers, rubbing them first from head towards tail with the sponge until they are well softened, when they can be treated more vigorously. Repeat this until all dirt appears to be removed.

Continue to immerse and wash the bird, then squeeze out all the soap you can and rinse in bowl No. 2. This water should also be fairly warm, and birds should be rinsed thoroughly, opening up the feathers (especially wings and hackle) to allow water freely through them. If soap remains in the feather the bird's last condition will be worse than the first.

Finally, rinse again in bowl No. 3, which I prefer to be warm, not tepid as usually recommended. The hotter the water the easier the drying process. A bird immersed in cold water dries miserably and slowly.

This final rinse should have a little blue added, but not much; an over-blued bird looks hopeless. In washing barred bantams, Silver Sebrights, etc., blue should be added; but colouring matter must not be used for breeds with ground colour other than white, because this is classed as faking.

Extra good results can be got by one more rinsing bowl, or by final rinsing under a warm-water tap. Then swab off as much surplus water as possible with the sponge. The bird can in fact be half-dried by sponging. Finally wipe it as dry as possible with a towel and put into a drying box or exhibition hamper before the fire.

When drying with a towel always start by wiping the wings inside and out, then under the wings and belly; and remember always to use the towel the way the feather lies, not against the web.

Most beginners in bantam keeping have few facilities, and in describing these processes it is assumed that electric hair-driers and similar aids are not available. However, if they are, all the better.

Those who exhibit regularly should make proper drying-boxes, but these can readily be improvised if necessary. Any box with open front covered by curtain-netting and suitable for standing before a fire will do. It is however better to make special boxes.

CHANGE THE POSITION

A number of boxes can be grouped around a fire, about two feet away, and birds should be turned round and not allowed constantly to stand in one position. When almost dry birds will preen themselves, and with their beaks transfer oil to their feathers from the oil gland at root of tail. If you are washing a lot of birds you can double-up birds in each compartment (cock and hen) when they are half dried.

Birds should be washed a couple of days before the show, and it is amazing how washing tames them. They will sit in the hand and talk to you after the first couple of times.

TRAVELLING HAMPERS

Don't send birds to show the day after washing. They aren't properly webbed out, and may take a chill; but don't be frightened of the job all the same—it's easy. Those who refrain from keeping white breeds merely on account of the washing involved miss a great sense of achievement.

It is not good to send birds to shows in flimsy structures, unlined and unsafe under pressure, so buy some proper travelling baskets, comfortably lined. In severe weather an additional outer fabric lining is advisable—and a waterproof cover for the lid.

If a number of birds are being sent, use hampers with partitions; and all hampers should have inner lids to each compartment, and a card-pocket under the main lid. White birds often get badly stained with colour from prize-cards carelessly dropped into their compartments.

Put sawdust, shavings or hay in the baskets. Probably the best virtue of hay is its cushioning power when hampers are roughly handled. Not all railway porters (or show stewards) are experienced with live-stock.

On its return from show, give your bird a good feed and a drink, whether you are pleased with results or not. If it has won it deserves credit; if it hasn't, it's likely your own fault. If it isn't well cared for it will feel strain after long journeys and possible exposure on draughty railway stations.

In any case, keep birds isolated after return for a day or two, to make sure they are healthy before turning them down with others. It is very annoying for birds to return with some contagious disease; but it's your fault if you're careless enough to let it spread through your flock. Luckily, nowadays disease isn't often transmitted through shows, but there's always the possibility; and sickness contracted from drinking cups may be of virulent character.

KEEPING BIRDS IN CONDITION

Show preparation doesn't end with pen-training and washing. We have to consider the periods between shows, when birds often need to be given outdoor conditions while at the same time kept clean to avoid too frequent tubbing.

After the first few washes birds can often go to a couple of shows on one wash; and although it is not advisable to send birds out constantly throughout a long show season, bantams are much more amenable in this respect than large fowl. They can be kept in show condition over protracted periods without undue loss of form or fitness; and when white birds are being sent frequently it is wise not to wash them every time if avoidable.

For this reason outdoor conditioning-coops are needed, to avoid penning show birds too long in indoor training cages. Some of these conditioning coops may resemble outdoor show-cages, suitable for use as hospital pens when required. Others may be simple box-form coops with open wire fronts; while yet more can be rather elaborate, taking the form of sheltered coops and runs permitting outdoor exercise but giving complete shade from sun, rain and wind.

CHAPTER 13

PARASITES AND DISEASES

AVOIDANCE IS BEST CURE

MANY people imagine that because bantams are tiny they will be subject to a great deal of sickness; but in actual fact they present few difficulties if sensibly maintained. Tiny size is not now regarded as important; and the development of genetic science and veterinary knowledge has made it possible to treat bantams almost as easily as large fowl.

There is little excuse for disease in household flocks— amongst which bantams are numbered; but be prepared in advance by keeping a small stock of remedies, and having a set of isolation cages—preferably of outdoor character and capable of easy cleaning.

When (or if) you get serious epidemics, or unexplained deaths, pay a fee and get a post-mortem examination, using the services of experts. This is much better than guessing and worrying.

In the light of recent knowledge, any full treatise on poultry diseases would need to divide them into several groups comprising (a) contagious, (b) infectious, (c) parasitic, (d) nutritional, and (e) organic diseases.

In such a book as this, however, it is only necessary to describe normal troubles. Bantam breeders and domestic poultry keepers are not likely to experience severe epidemics—and if they did, the little written here would not help them much; while long descriptions of deadly diseases (human nature being what it is) would merely cause them to examine their birds and imagine them as suffering from every ailment listed. Fresh air, adequate housing, cleanliness, and a sensible diet are the best measures for avoiding health problems.

Some diseases are both contagious and infectious, others parasitic and infectious. A third group may be both contagious and nutritional. Their categories are given in this short glossary of poultry troubles.

Tuberculosis: This used to be known as "going light", and many were the quack remedies sold to cure it. It is a chronic contagious complaint spread by contaminated droppings. Mortality is mainly in adults. Symptoms are extreme emaciation and bloodless appearance. Post-mortem examination shows a great number of yellow-white nodules in an enlarged liver, with intestines similarly affected. No treatment is advised. Disinfect housing, improve ventilation and treat open runs with quicklime, resting the pens for a month.

Fowl Paralysis: Mainly affects adults or near-adults. Cause is not definitely known, and there is no treament. Symptoms are lameness, drooping wings, loss of condition, paralysis and prostration, with contracted leg and foot muscles, and rigidity in head and neck.

B.W.D. (Bacillary White Diarrhœa): More properly known as pullorum disease. Caused by a bacillus present in the ovaries and transmitted through the egg. Affects baby chicks, and mortality is very heavy at 3–6 days. Adults may be carriers without visible signs. As a preventive measure, breeding stock may be blood-tested on the site and reactors destroyed. Symptoms—drowsiness, swaying and huddling, with poor appetite and very rapid deaths. Destruction of the whole batch is best. New drugs enable the disease to be controlled, but many fanciers may prefer to eliminate affected stock.

Coccidiosis: A parasitic disease that has several forms, chiefly affecting chicks at 3–6 weeks, with very heavy mortality. Symptoms are diarrhœa, usually with blood in droppings, unthriftiness, huddling and swaying; very similar to B.W.D. (which occurs however at a much younger age). Modern drugs are available from a veterinary surgeon which can cut losses to a minimum. Adult birds may be carriers.

Fowl Pox: Does not usually occur in domestic flocks, but was once common on farms. Also known as diphtheritic roup and canker. Caused by a virus, and symptoms are—scabs on face and comb, cheesy growths in mouth, and mucous discharge from nose and eyes. No treatment advised other than vaccination by veterinary surgeon. Kill affected birds and isolate contacts.

Favus: Often called white face. A fungoid parasitic disease that attacks comb and face, sometimes spreading to neck and body. Used to be extremely prevalent (especially in the North) but is now not often seen. Care is necessary in treatment, as humans may be affected. Symptoms are yellowish-white crusty spots on comb and face, spreading rapidly and giving a typical mouldy smell. Treat quickly by dressing with iodine, applying vaseline afterwards as an emollient. Apply several times at two-day intervals, and burn out infected coops with a blowlamp.

Fowl Pest: Sometimes called Newcastle disease. Is notifiable and highly infectious. In its acute form mortality is 100 per cent and very rapid. Symptoms are drowsiness, loss of appetite, unthriftiness, rapid breathing, rattling or gurgling, and twitching of head and neck; usually with partial paralysis of wings and legs and a strong mucous discharge from eyes and nostrils. A vaccine is now available which can keep the disease at bay.

Catarrh: Really a development of a severe cold, with sneezing, coughing and strong discharge from nostrils and eyes. Sometimes known as white-eye roup, and was always called roup by old-time poultry farmers. Highly infectious. Treatment—isolation and swabbing of nose and mouth with antiseptic and curative oils; but be sure to correct faulty ventilation and housing. Anti-biotic drugs may be employed, but are best administered by a veterinary surgeon.

Vent Gleet: More properly called Cloacitis. Venereal in character, and usually found only in breeding pens, where it spreads by copulation. Symptoms—inflamed vent with constant, smelly mucous discharge. Highly contagious. Treatment—wash frequently with good disinfectant and dress with antiseptic ointment. The safest remedy is to kill affected birds unless valuable.

Leukæmia: A form of anæmia, distinguishable only by post-mortem examination, therefore no treatment is advisable. Tumorous by nature.

Rickets: Most often due to deficiency of vitamin D, and normally attacks young chicks at about three weeks old—usually when reared behind ordinary glass indoors. Symptoms—swollen joints, doubled-up feet, crippled legs and buckled ribs. Treatment—direct daylight and 2 per cent cod liver oil in food.

Curly-toe lameness is another form of rickets. Add 2 per cent yeast to the food.

Slipped tendon: Properly called perosis, caused either by deficiency of manganese or excess phosphorus in food. Treatment not advised, but preventive measures include remedying the nutritional defects mentioned.

Prolapsus: An internal breakdown involving the egg-production organs. Mainly due to over-feeding and over-production, and aggravated by double-yolked and shell-less eggs. Treatment not advisable—kill and use for table while well-fleshed. Restrict diet in rest of flock.

(Protruding vent, dropped abdomen, down-behind, internal laying, water-belly, and all similar troubles are due to the same cause, and are all included under this general heading. The only hope of cure is semi-starvation for a week to check laying.)

Liver disease: Almost always due to over-feeding, and is common in heavily-fed layers. Symptoms—frothy, yellow, sloppy droppings, dark combs and lameness. Treatment should be semi-starvation for about four days, with a little liver pill each day. Then rest a couple of days and repeat. Should be killed for table unless a cure then results.

Peritonitis: Inflammation usually due to a broken egg in the oviduct. Seldom possible to treat, and only diagnosed by post-mortem examination.

Apoplexy: Sometimes called staggers. Usually due to brain congestion, and aggravated by hot weather. Treatment involves feeding very lightly, keeping in semi-darkness and dosing with Epsom Salts. However, often the bird is best killed.

Blood blisters mostly occur on thighs and breast and resemble small bladders of blood which burst and bleed severely. Cured by cauterising, but often break out again elsewhere. If troublesome, kill for table.

Nephritis is severe inflammation of the kidneys, common in poultry and usually caused by chill. No treatment is possible because it can only be diagnosed by post-mortem test.

Crop-binding is not the same as sour crop. In crop-binding food is held up, the entrance to the gullet (at base of crop) being choked by fibrous or stringy matter. It may be relieved by pouring oil down the throat and dispersing the obstruction by external massage. If this is not successful, it may be cured by a very simple operation, which any poultryman can undertake. Sour crop is a soured condition due to fermenting food, usually caused by an impacted gizzard. The crop may be emptied through the mouth, and generous dosing with medicinal paraffin will sometimes remove the impaction.

Impacted gizzard is caused by the gizzard being filled with more fibrous matter than it can cope with. The gizzard becomes choked and is out of action. Usually causes sour crop, hence is easy to diagnose. Medicinal paraffin through the mouth encourages clearance of the impaction, but if cure is not rapid, kill for table.

Poisoning is frequently the result of strong doses of common salt (sometimes through carelessness in using house-scraps) but is often due to picking up rat poison. The damage in either case is usually done before symptoms show, and thus treatment is useless.

Bronchitis is often the same thing as pneumonia, or accompanies it. Symptoms include heavy thirst and very strong breathing. Mild cases may be treated by dabbing curative oils (such as eucalyptus, camphor, etc.) in mouth and nostrils, but do not be too hopeful of cure. For valuable birds a veterinary surgeon should be consulted.

Bumble-foot usually results from bruising or cutting the foot. Seldom seen in bantams. Treat by lancing, removing pus, poulticing and isolating.

Worms affecting poultry are mainly roundworms, small caecal worms and tapeworms. They may be expelled by dosing with worm capsules obtainable from a veterinary surgeon.

Red mites do not live on the bodies of your birds, no matter what you may think. They are tiny grey mites that live in the crevices of houses. They visit birds in darkness, and are only red when full of the blood they suck. Treatment is to paint perch sockets and all crevices (or perhaps the whole house internally) with creosote 1 part, paraffin 2 parts; and use nicotine perch paint an hour before roosting-time.

If you see multitudes of tiny purple-red mites on your birds, often raising scabs under which they live, these are *not* red mite. They are known as Northern mite, and are best treated by washing the birds and disinfecting. They are much more vicious and difficult than red mite.

Scaly-leg is caused by an insect which burrows under scales and throws up a crusty deposit. Sulphur ointment is a good remedy, but in severe cases paint with a mixture of creosote and paraffin.

Depluming scabies are mites which burrow under the skin and destroy feathers, inflaming the skin, usually on head and neck. Treat by sulphur ointment, or by an emulsion of warm water, soft soap and sulphur, applied by well wetting the skin.

Lice and fleas may all be prevented and destroyed by the methods outlined for red-mite and Northern mite; also by dusting birds with a preparatory dusting powder.

Head ticks attack young birds, living with their heads buried under the scalp and causing many deaths. Carbolised vaseline (or carbolated oil) dabbed on the chicks' heads is death to the ticks.

You will observe that in nearly all severe cases of disease treatment is not only not advised—it is often impracticable; but you can prevent most diseases and parasites by cleanliness, and by frequently handling your stock. Prevention is far better than cure—especially with poultry, which have insufficient intelligence to assist treatment.

Worst causes of trouble are bad ventilation, lack of fresh air, and over-feeding. Nearly all organic troubles arise from too much food.

When you find your birds have crooked breasts, don't blame early perching. Blame intensive rearing for too long a period, lack of vitamins, or other nutritional deficiencies due to your own faulty observation.

Finally don't be frightened by these notes on parasites and disease. Good management is the best prevention. But in case you should experience sickness or epidemics, take care to have ready a small set of hospital pens, or at least to have an odd corner where one or two birds may be isolated in conditioning coops or outdoor cages.

Special Note: Tremendous advances have been made in recent years in developing new drugs. In the case of valuable birds where there is a serious problem with disease or infection, a veterinary surgeon should be consulted.

GLOSSARY OF TECHNICAL TERMS

A.O.C.: Any other Colour.

A.O.V.: Any other Variety.

Atavism: Reverison to ancestral type.

Bantam: Miniature fowl, formerly accepted as one-fifth the weight of the large breed it represented; but nowadays about one-fourth.

Barring: Alternate bands of dark and light colour across a feather, as in Plymouth Rocks and Cuckoo varieties.

Bay: A reddish-brown colour.

Beard: A tuft of feathers on the throat under the beak, in such varieties as Belgians, Houdans and Polands.

Beetle Brows: Heavy overhanging eyebrows, as in Malays and other Asiatic breeds.

Blade: The rear part of a single comb.

Blocky: Heavy and square in build.

Booted: Feathered on shanks and toes.

Brassiness: Yellowish foul colouring on plumage, usually on back and wing.

Breast: In live birds, the front of the body from keelbone to base of neck. In dead birds, the flesh on the keelbone.

Cap: The back part of a fowl's skull. (Refers mainly to colour.)

Cape: The feathers at base of neck hackle, covering the shoulders.

Carriage: General deportment or bearing, especially when walking.

Chicks: Young birds recently hatched.

Chicken: In exhibitions, a bird of the current season's breeding.

Cinnamon: A dark reddish-buff colour.

Cobby: Short-coupled and round or compact in build.

Cock: A yearling male bird or older.

Cockerel: A male bird of the current year's breeding.

Cockerel-bred: Bred in line from matings specially designed to produce good exhibition cockerels. (Compare with pullet-bred.)

Coverts: The covering feathers on tail and wing.

Crest: A tuft of feathers on top of head.

Cushion: A mass of soft feathers covering root of tail, as in Cochins or Pekins.

Cushion Comb: An almost hemispherical comb, as in the Silkie. Sometimes called strawberry comb.

Daw-eyed: Having pearl-coloured eyes.

Diamond: The wing bay.

Double-comb: An expression usually meaning rose-comb.

Dubbing: Removal (by cutting) of comb, wattles and ear-lobes.

Duck-foot: Having the rear toe lying close to the foot instead of spread out, thus resembling the foot of a duck.

137

Earlobes: Folds of skin below the ears proper; sometimes called deaf ears.

Flat Shin: Flat fronts to shanks—a serious defect in most breeds.

Flights: Primary wing feathers, unseen when wing is closed.

Fluff: The lower or downy part of a feather; the soft fluffy feather on thighs of soft-plumaged breeds.

Foul Feathers: Mismarked or wrongly-coloured feathers.

Foxy: Rusty or reddish in colour.

Gay: Light or white in markings of plumage.

Gypsy-face: A face dark, purple or mulberry in colour.

Ground-colour: Main colour of body plumage, on which markings are applied.

Hackles: The narrow pointed feathers of neck and saddle, particularly in males.

Hangers: The lesser sickle feathers and tail coverts.

Hard feather: Close tight feathers as found on Game birds.

Hen-feathered: A male bird without sickles or pointed hackles (sometimes called a henny).

Hock: The joint between thigh and shank.

Horn comb: A comb with two V-shaped horns.

Keel: The breastbone, particularly its edge.

Knock-kneed: In at the knees.

Lacing: An edging round the margin of a feather—usually single; but when double the outer lacing is always round the margin.

Leader: The spike at rear of a rose comb.

Leaf comb: Shaped like a butterfly, or like two leaves, one each side of the head.

Lopped comb: A comb falling over to one side.

Mealy: Stippled with lighter colour.

Moons: Rounded tips or spangles to feathers.

Mossy: Blurred or indistinct.

Mottled: Marked with tips or spots of different colour.

Muff: Feathers on each side of the face (often accompanying a beard); sometimes known as whiskers.

Muffling: Beard and whiskers enclosing the face.

Mulberry-faced: Gypsy-faced.

Pea-comb: A small triple comb with three longitudinal ridges, as in the Brahma.

Pearl-eyed: Sometimes called daw-eyed; eyes pearl-coloured.

Pencilling: This has three forms: (*a*) concentric rings of fine markings following the outline of the feather (but not round its edge) as seen in Partridge Wyandottes; (*b*) narrow barring in Pencilled Hamburghs—the only breed in which this is called pencilling; and (*c*) the fine stippled markings found on females of Brown Leghorns and Partridge O.E. Game, etc.

Primaries: Main flight feathers.

Pullet: A female fowl of current season's breeding.

Pullet-bred: Line-bred from a mating calculated to produce good exhibition females. (See cockerel-bred.)

Reachy: Tall and of upright carriage and "lift", as in Modern Game.

Roach-back: Humped back.

Rose-comb: A broad comb with small spikes or "work" on top, finishing with a spike or leader at rear. Sometimes called a double-comb.

Rusty: Reddish-brown or foxy in colour.

Saddle: The rear part of the back, adjoining the tail. Covered by the cushion in the hen.

Scaly-leg: A defective coral-like condition of the leg, caused by an insect parasite.

Secondaries: The second group of main quill feathers on wing adjoining the primaries. Unlike the primaries, visible when wing is closed.

Self-colour: One uniform colour, unmixed with other colours or markings.

Serrations: "Saw-tooth" sections of a single comb.

Shafty: Lighter on stem of feather than the ground-colour.

Sheen: Bright surface gloss on black plumage. In other colours usually described as lustre.

Shoulder: The upper part of the wing nearest the neck-feather. Prominent in Game breeds, where it is often called the shoulder-butt.

Sickles: Long curved tail feathers on a male bird.

Side sprig: An extra spike growing out of the side of a single comb.

Single-comb: A single fleshy growth or blade on head.

Slipped wing: A wing in which the primary flight feathers hang below the secondaries when wing is folded; a condition often allied with split wing, in which the primaries and secondaries show a very distinct segregation in many breeds of bantams.

Smut: Dark or smutty markings where undesirable, such as in under-colour of R.I.R.

Spangling: Markings (usually round) on the tips of feathers.

Spike: The rear leader on a rose comb.

Spur: A horny growth on the inside of a bird's shank, pronounced in cocks and used as a weapon of offence.

Squirrel-tail: A tail projecting forward towards the neck.

Strawberry-comb: A lump comb resembling half a strawberry, as in the Malay.

Striping: The very important markings down the middle of hackle feathers, particularly in males of Partridge varieties.

Sword-feathered: Having sickles only slightly curved, or scimitar-shaped, as in Japanese.

Symmetry: Proportion; harmony in shape.

Tail-coverts: See coverts.

Thumb-mark: A faulty indentation on the side of a single comb.

139

Ticked: Plumage tipped with different colour; usually applied to V-markings as in the Ancona.

Tri-coloured: A fault in buff birds, the hackles, wings, body-colour and tail showing three different shades.

Trio: A male and two females.

Type: Mould or shape; see symmetry.

Under-colour: Colour of fluff beneath surface plumage, not seen until the feather is lifted.

Variety: A sub-division of an established breed. It is considered that type makes the breed, colour and markings the variety.

Vulture hocks: Long stiff quill feathers projecting backwards from the hocks.

Wattles: Fleshy appendages below the beak, strongly developed in most male birds.

Wing-bar: A band of dark colour or markings across the middle of wing.

Wing-bay: The triangular surface of secondary feathers as exposed when wing is folded.

Wing-bow: The part of wing between wing-bar and shoulder.

Wing-butt: The end of the wing; end of primaries.

Wing-coverts: Feathers covering roots of secondary feathers.

Work: The small spikes or working on top of a rose comb.

Wry-tail: A defective tail carried habitually to one side.

INDEX OF BREEDS

J. D. KAY

Breeder, Exhibitor and Exporter of :-

SILVER SPANGLED HAMBURGHS
(Large and Bantams)
BLACK HAMBURGHS
GOLD and SILVER SEBRIGHTS
BLACK ROSECOMBS

Winner of Hundreds of First Prizes and Many Championships

All Enquiries to:—
LITTLE OXENDALE FARM
OSBALDESTON, Nr. BLACKBURN, LANCS
Telephone Mellor 2883

KAY — ONE OF THE GREAT NAMES IN EXHIBITION POULTRY

THE POULTRY CLUB

Caters for anyone who is interested in Poultry for Pleasure or Exhibition

Membership gives many advantages including a free copy of the Poultry Club Year Book as well as contact with others in the fancy

Secretary :
Mrs. S. JONES, 72 SPRINGFIELDS, DUNMOW, ESSEX